Study Guide with Solutions to Selected Problems

Essentials of General, Organic, and Biological Chemistry

H. Stephen Stoker

Danny V. White

Joanne A. White

Houghton Mifflin Company   Boston   New York

Editor-in-Chief: Charles Hartford
Executive Editor: Richard Stratton
Developmental Editor: Rita Lombard
Editorial Assistant: Rosemary Mack
Senior Manufacturing Coordinator: Florence Cadran
Executive Marketing Manager: Katherine Greig

Printed in the U.S.A.

ISBN: 0-618-192832

56789- DBH- 06

Contents

Preface   v

Chapter 1   **Basic Concepts About Matter**
Study Guide   1
Solutions   173

Chapter 2   **Measurements in Chemistry**
Study Guide   8
Solutions   175

Chapter 3   **Atomic Structure, the Periodic Law, and Radioactivity**
Study Guide   18
Solutions   177

Chapter 4   **Chemical Bonds**
Study Guide   25
Solutions   180

Chapter 5   **Chemical Calculations: Formula Masses, Moles, and Chemical Equations**
Study Guide   36
Solutions   183

Chapter 6   **Gases, Liquids, and Solids**
Study Guide   45
Solutions   186

Chapter 7   **Solutions**
Study Guide   54
Solutions   189

Chapter 8   **Chemical Reactions**
Study Guide   63
Solutions   192

Chapter 9   **Acids, Bases, and Salts**
Study Guide   71
Solutions   194

Chapter 10   **Saturated Hydrocarbons**
Study Guide   80
Solutions   196

Chapter 11   **Unsaturated Hydrocarbons**
Study Guide   91
Solutions   199

Chapter 12   **Hydrocarbon Derivatives I: Carbon Heteroatom Single Bonds**
Study Guide   103
Solutions   202

Chapter 13   **Hydrocarbon Derivatives II:
Carbon Oxygen Double Bonds**
                 Study Guide   115
                 Solutions   206
Chapter 14   **Carbohydrates**
                 Study Guide   130
                 Solutions   210
Chapter 15   **Lipids**
                 Study Guide   138
                 Solutions   213
Chapter 16   **Proteins**
                 Study Guide   146
                 Solutions   216
Chapter 17   **Nucleic Acids**
                 Study Guide   155
                 Solutions   219
Chapter 18   **Metabolism**
                 Study Guide   163
                 Solutions   222

Preface

If the study of chemistry is new to you, you are about to gain a new perspective on the material world. You will never again look at the objects and substances around you in quite the same way. The invisible structure and organization of matter, the "how and why" of chemical change -- knowledge of these areas will help to demystify many occurrences in the world around you. Chemistry is not an isolated academic study. We use it throughout our lives to appreciate and understand the world and to make responsible choices in that world.

The purpose of this study guide is to help you in your study of the textbook, *Essentials of General, Organic, and Biological Chemistry*, by providing summaries of the text and additional practice exercises. As you use this Study Guide, we suggest that you follow the steps below.

1. Read the overview for the chapter to get a general idea of the facts and concepts in each chapter.

2. Read the section summaries and work the practice exercises as you come to them. Write out the answers even if you are sure you understand the concepts. This will help you to check your understanding of the material. Refer to the answers at the end of the chapter as soon as you have answered each practice exercise. By checking your answers, you will know whether to review or continue to the next section.

3. When you have finished answering the practice exercises, take the self-test at the end of the chapter. Check your answers with the answer key at the end of the chapter. If there are any questions that you answer incorrectly or do not understand, refer to the chapter section numbers in the answer key and review that material. The Solutions section of this book contains answers to selected exercises and problems from the textbook along with methods for solving many of the problems.

Chemistry is a discipline of patterns and rules. Once your mind has begun to understand and accept these patterns, the time you have spent on repetition and review will be well rewarded by a deeper total picture of the world around you. As teachers, we have enjoyed preparing this study guide and hope that it will assist you in your study of chemistry.

Danny and Joanne White

Basic Concepts About Matter   Chapter 1

Chapter Overview

Why is the study of chemistry important to you? Chemistry produces many substances of practical importance to us all: building materials, foods, medicines. For anyone entering one of the life sciences, such as the health sciences, agriculture or forestry, an understanding of chemistry leads to an understanding of the many life processes.

In this chapter you will be studying some of the fundamental ideas and the language of chemistry. You will characterize three states of matter, differentiate between physical properties and chemical properties, and identify two different types of mixtures. You will describe elements and compounds, and practice using symbols and formulas.

Practice Exercises

1.1   **Matter** (Sec. 1.1) exists in three physical states. Complete the following table indicating the properties of each of these states of matter.

State	Definite shape?	Definite volume?
solid (Sec. 1.2)	yes	
liquid (Sec. 1.2)		
gas (Sec. 1.2)		

1.2   The **physical properties** (Sec. 1.3) of a substance can be observed without changing the identity of the substance. **Chemical properties** (Sec. 1.3) are observed when a substance changes or resists changing to another substance. Complete the following table:

Property	Physical	Chemical	Insufficient information
liquid boils at 100°C			
solid forms a gas when heated			
metallic solid exposed to air forms a white solid			

1.3   A **physical change** (Sec. 1.4) is a change in shape or form, but not in composition. A **chemical change** (Sec. 1.4) produces a new substance; that is, the composition is changed. Classify the following processes by marking the correct column:

Process	Physical change	Chemical change
ice cube melts		
wood block burns		
salad oil freezes solid		
sugar dissolves in tea		
wood block is split		
butter becomes rancid		

1

1.4   **Mixtures** (Sec. 1.5) of substances may be either **homogeneous** (Sec. 1.5), one phase (part), uniform throughout, or **heterogeneous** (Sec. 1.5), visibly different parts or phases. Indicate whether each of the following mixtures is homogeneous or heterogeneous:

Mixture	Homogeneous	Heterogeneous
apple juice (water, sugar, fruit juice)		
cornflakes and milk		
fruit salad		
brass (copper and zinc)		

1.5   **Elements** (Sec. 1.6) are **pure substances** (Sec. 1.5) that cannot be broken down into simpler pure substances. **Compounds** (Sec. 1.6) can be broken down into two or more simpler pure substances by chemical means. Complete the following table:

	Substance is an element	Substance is a compound	Insufficient information to classify
Substance A reacts violently with water.			
Substance B can be broken into simpler substances by chemical processes			
Cooling substance C at 350°C turns it from a liquid to a solid.			
Substance D cannot decompose into simpler substances by chemical processes.			

1.6   In the following table, write the **chemical symbol** (Sec. 1.8) or name for each element:

Name	Symbol
calcium	
copper	
argon	
nickel	
magnesium	

Symbol	Name
C	
Ne	
Zr	
Pb	
Fe	

1.7   Complete the following diagram:

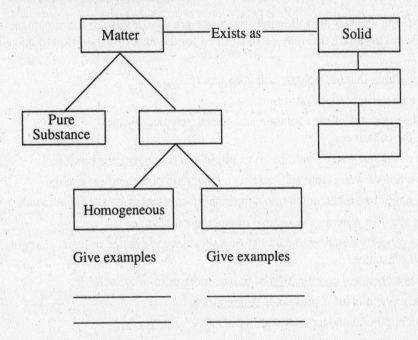

Give examples   Give examples

_____   _____

_____   _____

1.8   An **atom** (Sec. 1.9) is the smallest particle of an element that keeps the identity of that element. A **molecule** (Sec. 1.9) is a tightly-bound group of two or more atoms that functions as a unit. **Homoatomic molecules** (Sec. 1.9) are made up of atoms of one element; **heteroatomic molecules** (Sec. 1.9) contain atoms of two or more elements.

In the table below, indicate whether each unit of a substance is an atom, a homoatomic molecule, or a heteroatomic molecule, and classify the substance as an element or a compound:

Unit	Atom	Homoatomic molecule	Heteroatomic molecule	Element	Compound
N_2					
CH_4					
HCN					
Au					
HF					

1.9   Write a **chemical formula** (Sec. 1.10) for each of the following compounds based on the information given:

a.   A molecule of limonene contains 10 atoms of carbon and 16 atoms of hydrogen.

b.   A molecule of nitric acid is pentatomic (Sec. 1.9) and contains the elements hydrogen, nitrogen and oxygen. Each molecule of nitric acid contains only one atom of hydrogen and one of nitrogen.

Self-Test

True-false: Indicate whether the following statements are true or false. If the statement is false, give the word or phrase that may be substituted for the underlined portion to make the statement true.

1. Matter is anything that has <u>volume</u> and occupies space.

2. <u>Gases</u> have no definite shape or volume.

3. <u>Liquids</u> take the shape of the container and completely occupy the volume of the container.

4. A mixture of oil and water would be an example of a <u>homogeneous</u> mixture.

5. The evaporation of water from salt water is an example of a <u>chemical change</u>.

6. Elements <u>cannot</u> be broken down into simpler pure substances by chemical means.

7. Dissolving sugar in water is an example of a <u>chemical change</u>.

8. A <u>chemical property</u> describes the ability of a substance to change or resist changing to form a new substance.

9. A mixture is a <u>chemical combination</u> of two or more pure substances.

10. Most of the oxygen in air is present as <u>single atoms</u>.

11. The most common (abundant) element in the universe is <u>oxygen</u>.

12. Two-letter chemical symbols are <u>always</u> the first two letters of the element's name.

13. A compound consists of molecules that are <u>homoatomic</u>.

Multiple choice:

14. One of the three states of matter is a solid. A solid has the following characteristics:

   a. definite volume, no definite shape   b. no definite volume, no definite shape
   c. definite volume and shape   d. no definite volume, but definite shape
   e. none of these

15. An example of a homogeneous mixture would be:

   a. sand and water   b. salt and water   c. wood and water
   d. oil and water   e. none of these

16. An example of a physical change would be:

   a. rusting of iron   b. sugar dissolving in coffee
   c. gasoline burning in a car engine   d. burning coal
   e. none of these

17. An example of a chemical change would be:

   a. rusting of iron   b. burning coal
   c. gasoline burning in a car engine   d. a, b, and c are all correct
   e. none of these

18. The symbol for the element iron is:

   a. FE   b. Fe   c. F   d. Ir   e. none of these

19. The name for the element Ne is:

   a. neodymium   b. neon   c. neptunium
   d. nitrogen   e. none of these

20. $MgCO_3$ is a compound that is composed of which elements?

   a. magnesium, chlorine, iron   b. magnesium, carbon, neon
   c. manganese, carbon, oxygen   d. magnesium, carbon, oxygen
   e. none of these

21. The total number of elements in the compounds sodium sulfate (Na_2SO_4) is

   a. 2   b. 3   c. 4   d. 5   e. 7

22. The total number of atoms in one molecule of CH_4O is:

   a 3   b. 4   c. 5   d. 6   e. none of these

23. On the basis of its formula, which of the following substances is an element?

   a. NH_3 b. Cl_2 c. CO_2 d. CO   e. none of these

24. On the basis of its formula, which of the following is a compound?

   a. Fe   b. Fm   c. O_2 d. HI   e. none of these

Answers to Practice Exercises

1.1

State	Definite shape?	Definite volume?
solid	yes	yes
liquid	no	yes
gas	no	no

1.2

Property	Physical	Chemical	Insufficient information
liquid boils at 100°C	X		
solid forms a gas when heated			X
metallic solid exposed to air forms a white solid		X	

1.3

Process	Physical change	Chemical change
ice cube melts	X	
wood block burns		X
salad oil freezes solid	X	
sugar dissolves in tea	X	
wood block is split	X	
butter becomes rancid		X

1.4

Mixture	Homogeneous	Heterogeneous
apple juice (water, sugar, fruit juice)	X	
cornflakes and milk		X
fruit salad		X
brass (copper and zinc)	X	

1.5

	Substance is an element	Substance is a compound	Insufficient information to classify
Substance A reacts violently with water.			X
Substance B can be broken into simpler substances by chemical processes.		X	
Cooling substance C at 350°C turns it from a liquid to a solid.			X
Substance D cannot decompose into simpler substances by chemical processes.	X		

1.6

Name	Symbol
calcium	Ca
copper	Cu
argon	Ar
nickel	Ni
magnesium	Mg

Symbol	Name
C	carbon
Ne	neon
Zr	zirconium
Pb	lead
Fe	iron

1.7

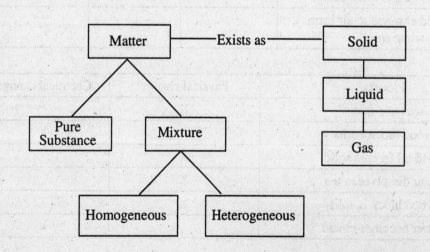

Give examples of homogeneous mixtures:

Give examples of heterogeneous mixtures:

air

smoke

coffee

concrete

(These are a few of many possible examples.)

1.8

Unit	Atom	Homoatomic molecule	Heteroatomic molecule	Element	Compound
N_2		X		X	
CH_4			X		X
HCN			X		X
Au	X			X	
HF			X		X

1.9 a. $C_{10}H_{16}$ b. HNO_3

Answers to Self-Test

The numbers in parentheses refer to sections in your textbook.
1. F; mass (1.1) **2.** T (1.2) **3.** F; gases (1.2) **4.** F; heterogeneous (1.5) **5.** F; physical change (1.4)
6. T (1.6) **7.** F; physical change (1.4) **8.** T (1.3) **9.** F; physical combination (1.5)
10. F; diatomic molecules (1.9) **11.** F; hydrogen (1.7) **12.** F; often (1.8) **13.** F; heteroatomic (1.9)
14. c (1.2) **15.** b (1.5) **16.** b (1.4) **17.** d (1.4) **18.** b (1.8) **19.** b (1.8) **20.** d (1.8) **21.** b (1.10)
22. d (1.10) **23.** b (1.6, 1.10) **24.** d (1.10)

Chapter Overview

Measurements are very important in science. In this chapter you will study some of the common units used in measuring length, volume, mass, temperature, and heat in the metric system. Numbers resulting from measurements are always inexact, so you will practice using the number of significant figures that corresponds to the uncertainty of the measurements being made.

You will solve problems involving measurements using the method of dimensional analysis, in which units associated with numbers are used as a guide in setting up the calculations, and use scientific notation to express large and small numbers efficiently. You will also use equations to calculate density and to convert temperature from one temperature scale to another.

Practice Exercises

2.1   **Exact numbers** (Sec. 2.3) occur in definitions, in counting, and in simple fractions. **Inexact numbers** (Sec. 2.3) are obtained from measurements. Classify the numbers in the following statements as exact or inexact by marking the correct column.

Number	Exact	Inexact
A bag of sugar weighs 5 pounds.		
The temperature was 104°F in the shade.		
There were 107 people in the airplane.		
An octagon has eight sides.		
One meter equals 100 centimeters.		
1/3 is a simple fraction.		
The swimming pool was 25 meters long.		

2.2   **Significant figures** (Sec. 2.4) are the digits in any **measurement** (Sec. 2.1) that are known with certainty plus one digit that is uncertain. These guidelines will help you in determining the number of significant figures:

1. All nonzero digits are significant. (23.4 m has three significant figures)
2. Zeros in front of nonzero digits are not significant. (0.00025 has two significant figures)
3. Zeros between nonzero digits are significant. (2.005 has four significant figures)
4. Zeros at the end of a number are significant if a decimal point is present (1.60 has three significant figures), but not significant if there is no decimal point (500 has one significant figure; 500. has three significant figures).

In **rounding off** (Sec. 2.5) a number to a certain number of significant figures:

1. Look at the first digit to be deleted.
2. If that digit is less than 5, drop that digit and all those to the right of it. If that digit is 5 or more, increase the last significant digit by one.

Example: Round 7.3589 to three significant figures. Since 7.35 has the correct number of significant figures and 8 is greater than 5, increase 7.35 to 7.36.

In the following table state the number of significant figures in each number as written, and then round off the number to 3 significant figures.

Number	Number of significant figures	Rounded to 3 significant figures
1578		
45932		
103045		
0.00034722		
0.0345047		
0.000450700		
2984400		

2.3   When multiplying and dividing measurements, the number of significant figures in the answer is the same as the number of significant figures in the measurement that contains the fewest significant figures (Sec. 2.5).

Example: 4.2 m x 3.12 m = 13.104 m^2 = 13 m^2
(Since 4.2 has fewer significant figures, the answer has two significant figures.)

Exact numbers (such as three people or twelve eggs in a dozen) do not limit the number of significant figures.

Carry out each of the mathematical operations as indicated below, and round the answer to the correct number of significant figures. (Assume there are no exact numbers in any calculation.):

Problem	Answer before rounding off	Rounded to correct number of significant figures
160 x 0.32		
482 x 0.00358		
5 x 3.00985		
723 ÷ 4.04		
72 ÷ 1.37		
0.0485 ÷ 88.342		

2.4   When measurements are added and subtracted, the answer can have no more digits to the right of the decimal point than the measurement with the least number of decimal places (Sec. 2.5).

Example: 3.58 m + 7.2 m = 10.78 m = 10.8 m

Carry out the mathematical operations as indicated below, and round the answer to the correct number of significant figures.

Problem	Answer before rounding off	Rounded to correct number of significant figures
153 + 4521		
483 + 0.223		
1.097 + 0.34		
744 − 36		
8093 − 0.566		
0.345 − 0.0221		

2.5 **Scientific notation** (Sec. 2.6) is a convenient way of expressing very large or very small numbers in a compact form. To convert a number to scientific notation:
1. Write the original number.
2. Move the decimal point to a position just to the right of the first nonzero digit.
3. Count the number of places the decimal point was moved. This number will be the exponent of 10.
4. If the decimal point is moved to the left, the exponent will be positive; if the decimal point is moved to the right the exponent will be negative.

Examples:
1. $2300 = 2.3 \times 10^3$ (decimal moved 3 places to the left)
2. $43,010,000 = 4.301 \times 10^7$ (decimal moved 7 places to the left)
3. $0.0072 = 7.2 \times 10^{-3}$ (decimal moved 3 places to the right)

Note that only significant figures become a part of the coefficient.

Complete the following tables:

Decimal number	Scientific notation
4378	
783	
8401	
0.00362	
0.093200	

Scientific notation	Decimal number
6.389×10^6	
3.34×10^1	
4.55×10^{-3}	
9.08×10^{-5}	
2.0200×10^{-2}	

2.6 Use the method of **dimensional analysis** (Sec. 2.7) to solve the problems below. Choose the **conversion factor** (Sec. 2.7) that gives the answer in the correct units. If the wrong conversion factor is used the units will not cancel.

Example: How many millimeters (mm) are in 2.41 meters (m)?

$$2.41 \text{ m} \times \frac{1000 \text{ mm}}{1 \text{ m}} = (2.41 \times 1000) \left(\frac{\text{m} \times \text{mm}}{\text{m}}\right) = 2410 \text{ mm}$$

Complete the following table. Use the factors in Table 2.2 of your textbook for conversion between the metric and English systems.

Problem	Relationship with units	Answer with units
3.89 km = ? cm	$3.89 \text{ km} \times \dfrac{1000 \text{ m}}{1 \text{ km}} \times \dfrac{100 \text{ cm}}{1 \text{ m}}$	
45.8 mm = ? m		
0.987 mm = ? km		
7.89×10^4 cm = ? m		
4.05×10^{-4} cm = ? km		
5.5×10^5 L = ? mL		
3.6 km = ? in.		
8.2 mL = ? qt		
57 fl oz = ? mL		
3.8 oz = ? kg		

2.7   **Density** (Sec. 2.8) is the ratio of the **mass** (Sec. 2.2) of an object to the volume occupied by that object.

Example: An object has a mass of 123 g and a volume of 17 cm³. Calculate the density of the object.

$$\text{Density} = \frac{\text{mass}}{\text{volume}} = \frac{123 \text{ g}}{17 \text{ cm}^3} = 7.2 \text{ g/cm}^3$$

Example: What mass will a cube of aluminum have if its volume is 32 cm³? Aluminum has a density of 2.7 g/cm³.

Using the method of dimensional analysis, we can set up an equation that uses the density as a conversion factor:

32 cm³ = ? g

$$32 \text{ cm}^3 \times \frac{2.7 \text{ g}}{1.0 \text{ cm}^3} = 86 \text{ g}$$

Complete the following table:

Problem	Substituted equation with units	Answer with units
1. What is the density of silver, if 38.5 cm³ has a mass of 404 grams?		
2. Ethanol has a density of 0.789 g/mL. What is the mass of 458 mL of ethanol?		
3. Copper has a density of 8.92 g/cm³. What would be the volume of 8.97 kg of copper?		

2.8   Convert the boiling point temperatures of the following compounds to the indicated temperature scales. Use the equations in Section 2.9 of your textbook.

Boiling point	Substituted Equation	Temperature
ethyl acetate (nail polish remover) 77.0°C	°F = 9/5(77.0) + 32	°F
toluene (additive in gasoline) 111°C		°F
isopropyl alcohol (rubbing alcohol) 180°F		°C
naphthalene (moth balls) 424°F		°C
propane (fuel for camping stoves) −42°C		K
methane (natural gas) −260°F		K

2.9   A calorie (Sec. 2.9) is the amount of heat energy needed to raise the temperature of 1 gram of water by 1 degree Celsius. One calorie equals 4.184 joules, and 1000 calories equals 1 kilocalorie (or one Calorie).

Example: It took 1.648 kilocalories to melt a block of ice. How many calories of heat energy is equivalent to 1.648 kilocalories? How many joules of heat energy?

1. 1 kilocalorie = 1000 calories

$$\text{heat (cal)} = 1.648 \text{ kcal} \times \frac{1000 \text{ cal}}{1 \text{ kcal}} = 1,648 \text{ cal}$$

2. 1 calorie = 4.184 joules

$$\text{heat (J)} = 1,648 \text{ cal} \times \frac{4.184 \text{ J}}{1 \text{ cal}} = 6895 \text{ J}$$

In each row of the table below, convert the given heat energy to its equivalent energy in the other two units.

	kilocalories	calories	joules
1.	2.143 kcal		
2.		518.8 cal	
3.			7652 J

Self-Test

True-false: Indicate whether the following statements are true or false. If the statement is false, give the word or phrase that may be substituted for the underlined portion to make the statement true.

1. In the metric unit system the base unit of volume is the <u>milliliter</u>.

2. The base unit of mass in the metric system is the <u>kilogram</u>.

3. In scientific notation, a <u>coefficient</u> between 1 and 10 is multiplied by a power of ten.

4. The ease of conversion from one unit to another, by dividing or multiplying by multiples of ten, is an advantage of the <u>English system</u> of measurement.

5. In rounding numbers, the number of digits is determined by the number of <u>significant figures</u> in the measurement.

6. The density of an object is the ratio of <u>its weight to its volume</u>.

7. The calorie is a common measure of heat and is defined as the quantity of heat that raises the temperature of <u>1 gram of water 1°C</u>.

8. In the metric system, the prefix <u>micro-</u> means one-thousandth (0.001, 10^{-3})

9. The number, 0.002010, has <u>four</u> significant figures.

10. On the Kelvin temperature scale, <u>all</u> temperature readings are positive.

Multiple choice:

11. Which of the following statements contains an exact number?

    a. The Earth's moon has a diameter of 2160 miles.
    b. The temperature of the room is 68°F.
    c. There were 14 bananas in the bunch.
    d. The student completed the test in 40 minutes.
    e. none of these

12. The number of joules equal to 212 calories is:

    a. 887 joules   b. 50.7 joules   c. 0.887 joules
    d. 507 joules   e. 8.87 joules

13. The correct way of expressing 4174 in scientific notation is:

    a. 4.174×10^2 b. 4.174×10^3 c. 4.174×10^4
    d. 4.2×10^3 e. none of these

14. The number, 0.005140, should be written in scientific notation as:

    a. 5.140×10^{-2} b. 514×10^{-4} c. 5.14×10^{-3}

    d. 5.140×10^{-3} e. none of these

15. The sum of 5472 plus 1946 would be written in scientific notation as:

    a. 7418   b. 7.418×10^{3} c. 7.4×10^{3}

    d. 7.42×10^{3} e. none of these

16. The product of 8311 times 0.01452 would be written in scientific notation as:

    a. 120.7   b. 1.21×10^{2} c. 1.207×10^{2}

    d. 1.2×10^{2} e. none of these

17. When 1.487 is rounded to three significant figures, the correct answer is:

    a. 1.480   b. 1.490   c. 1.48   d. 1.49   e. none of these

18. In converting grams to milligrams, the known quantity of grams should be multiplied by which of these conversion factors?

    a. 1 g/1000 mg   b. 1000 mg/1 g   c. 1 g/1,000,000 µg

    d. 1,000,000 µg/1 g   e. none of these

19. What is the volume, in mL, occupied by 25.2 kilograms of a liquid whose density is 0.833 g/mL?

    a. 3.03×10^{4} mL   b. 21.0 mL   c. 2.10×10^{4} mL

    d. 30.3 mL   e. none of these

20. Which of the following measurements has three significant figures?

    a. 1.050   b. 2301 mL   c. 0.0702 g

    d. 16.20 m   e. none of these

21. A temperature of 22°C would have which of these values on the Fahrenheit scale?

    a. –6.6°F   b, 54°F   c. 97°F   d. 72°F   e. none of these

Answers to Practice Exercises

2.1

Number	Exact	Inexact
A bag of sugar weighs 5 pounds.		X
The temperature was 104°F in the shade.		X
There were 107 people in the airplane.	X	
An octagon has eight sides.	X	
One meter equals 100 centimeters.	X	
1/3 is a simple fraction.	X	
The swimming pool was 25 meters long.		X

2.2

Number	Number of significant figures	Rounded to 3 significant figures
1578	4	1580
45932	5	45900
103045	6	103000
0.00034722	5	0.000347
0.0345047	6	0.0345
0.000450700	6	0.000451
2984400	5	2980000

2.3

Problem	Answer before rounding off	Rounded to correct number of significant figures
160 x 0.32	51.2	51
482 x 0.00358	1.72556	1.73
5 x 3.00985	15.04925	20
723 ÷ 4.04	178.96039	179
72 ÷ 1.37	52.554744	53
0.0485 ÷ 88.342	0.000549	0.000549

2.4

Problem	Answer before rounding off	Rounded to correct number of significant figures
153 + 4521	4674	4674
483 + 0.223	483.223	483
1.097 + 0.34	1.437	1.44
744 − 36	708	708
8093 − 0.566	8092.434	8092
0.345 − 0.0221	0.3229	0.323

2.5

Decimal number	Scientific notation
4378	4.378×10^3
783	7.83×10^2
8401	8.401×10^3
0.00362	3.62×10^{-3}
0.093200	9.3200×10^{-2}

Scientific notation	Decimal number
6.389×10^6	6,389,000
3.34×10^1	33.4
4.55×10^{-3}	0.00455
9.08×10^{-5}	0.0000908
2.0200×10^{-2}	0.020200

2.6

Problem	Relationship with units	Answer with units
3.89 km = ? cm	3.89 km x $\dfrac{1000\ m}{1\ km}$ x $\dfrac{100\ cm}{1\ m}$	3.89×10^5 cm
45.8 mm = ? m	45.8 mm x $\dfrac{1\ m}{1000\ mm}$	4.58×10^{-2} m
0.987 mm = ? km	0.987 mm x $\dfrac{1\ m}{1000\ mm}$ x $\dfrac{1\ km}{1000\ m}$	9.87×10^{-7} km
7.89×10^4 cm = ? m	7.89×10^4 cm x $\dfrac{1\ m}{100\ cm}$	7.89×10^2 m
4.05×10^{-4} cm = ? km	4.05×10^{-4} cm x $\dfrac{1\ m}{100\ cm}$ x $\dfrac{1\ km}{1000\ m}$	4.05×10^{-9} km
5.5×10^5 L = ? mL	5.5×10^5 L x $\dfrac{1000\ mL}{1\ L}$	5.5×10^8 mL
3.6 km = ? in.	3.6 km x $\dfrac{1000\ m}{1\ km}$ x $\dfrac{39.4\ in.}{1.00\ m}$	1.4×10^5 in.
8.2 mL = ? qt	8.2 mL x $\dfrac{1\ L}{1000\ mL}$ x $\dfrac{1.00\ qt}{0.946\ L}$	8.7×10^{-3} qt
57 fl oz = ? mL	57 fl oz x $\dfrac{1.00\ mL}{0.034\ fl\ oz}$	1.7×10^3 mL
3.8 oz = ? kg	3.8 oz x $\dfrac{28.3\ g}{1.00\ oz}$ x $\dfrac{1\ kg}{1000\ g}$	1.1×10^{-1} kg

2.7

Problem	Substituted equation and answer
1. What is the density of silver, if 38.5 cm³ has a mass of 404 grams?	$\text{density} = \dfrac{\text{mass}}{\text{volume}} = \dfrac{404\ g}{38.5\ cm^3} = 10.5\ g/cm^3$
2. Ethanol has a density of 0.789 g/mL. What is the mass of 458 mL of ethanol?	Use density as a conversion factor. $\text{mass} = \dfrac{0.789\ g}{1\ mL} \times 458\ mL = 361\ g$
3. Copper has a density of 8.92 g/cm³. What would be the volume of 8.97 kg of copper?	Use two conversion factors to solve this proplem. $\text{volume} = \dfrac{1.00\ cm^3}{8.92\ g} \times \dfrac{1000\ g}{1\ kg} \times 8.97\ kg = 1.01 \times 10^3\ cm^3$

2.8

Boiling point	Substituted equation	Temperature
ethyl acetate (nail polish remover) 77°C	°F = 9/5(77.0) + 32	171°F
toluene (additive in gasoline) 111°C	°F = (9/5)111 + 32	232°F
isopropyl alcohol (rubbing alcohol) 180°F	°C = 5/9(180 − 32)	82°C
naphthalene (moth balls) 424°F	°C = 5/9(424 − 32)	218°C
propane (fuel for camping stoves) −42°C	K = (−42) + 273	231 K
methane (natural gas) −260°F	°C = 5/9(−260−32) = − 162°C K = (−162) + 273	111 K

2.9

	kilocalories	calories	joules
1.	2.143 kcal	2143 cal	8966 J
2.	0.5188 kcal	518.8 cal	2171 J
3.	1.829 kcal	1829 cal	7652 J

Answers to Self-Test

The numbers in parentheses refer to sections in your textbook:
1. F; liter (2.2) **2.** T (2.2) **3.** T (2.6) **4.** F; metric unit system (2.2) **5.** T (2.5)
6. F; its mass to its volume (2.8) **7.** T (2.9) **8.** F; milli- (2.2) **9.** T (2.3) **10.** T (2.8)
11. c (2.3) **12.** a (2.9) **13.** b (2.6) **14.** d (2.6) **15.** b (2.6) **16.** c (2.4 and 2.6)
17. d (2.5) **18.** b; (2.7) **19.** a (2.8) **20.** c (2.4) **21.** d (2.9)

Atomic Structure, the Periodic Law, and Radioactivity
Chapter 3

Chapter Overview

All matter is made of basic building blocks called atoms. As you study the structure of atoms, you can begin to develop an understanding of how atoms bond together to form the many substances which make up our world. By studying the periodic law, you will begin to predict the properties of elements according to their positions in the periodic table.

By the end of this chapter you should be able to describe the three basic particles which make up atoms in terms of mass, charge and location, and calculate the number of each of the three types of particles in an atom using the atomic number and the mass number of that atom. You will learn to describe an isotope from its symbol, and you will calculate the average atomic mass of an element. Using the electron configuration of an element and the principle of the distinguishing electron, you will be able to classify the elements into groups with similar properties.

In nuclear reactions, an atom's nucleus changes. You will compare three kinds of nuclear radiation and study the rate of radioactive decay of some isotopes.

Practice Exercises

3.1 The **atomic number** (Sec. 3.2) of an **element** (Sec. 3.2) is the number of **protons** (Sec. 3.1) in the **nucleus** (Sec. 3.1) of that element. Since the number of protons in an atom is equal to the number of **electrons** (Sec. 3.1), the atomic number also gives us the number of electrons in a neutral atom. For example, the atomic number of oxygen is 8; therefore, the oxygen atom has 8 protons and 8 electrons.

The **mass number** (Sec. 3.2) is the total number of **nucleons** (protons and **neutrons**) (Sec. 3.1) in the nucleus. Since the atomic number is equal to the number of protons in the nucleus, and the mass number is equal to protons plus neutrons, we can find the number of neutrons by subtraction: Mass number – Atomic number = Number of neutrons

Use the relationships above and the **periodic table** (Sec. 3.4) to complete the following table:

Atomic number	Mass number	Number of protons	Number of neutrons	Number of electrons	Symbol of element
6			6		
	39	19			
			77		Xe
	64			29	
		35	45		

3.2 **Isotopes** (Sec. 3.3) are atoms that have the same number of protons but different numbers of neutrons, and therefore different mass numbers. Isotopes are usually represented as follows:

$$^{14}_{6}C$$

The superscript is the mass number, or A.
The subscript is the atomic number, or Z.

Complete the following table:

Isotope	A	Z	Protons	Neutrons	Electrons
$^{40}_{20}$Ca					
$^{40}_{18}$Ar					
$^{23}_{11}$Na					
$^{37}_{17}$Cl					
$^{35}_{17}$Cl					

3.3 The **atomic mass** (Sec. 3.3) of an element is an average mass of the mixture of isotopes that reflects the relative abundance of the isotopes as they occur in nature. The atomic mass can be calculated by multiplying the relative mass of each isotope by its fractional abundance, and then totaling the products.

Example:
Magnesium is composed of 78.7% ^{24}Mg, 10.1% ^{25}Mg and 11.2% ^{26}Mg. To find the atomic mass for magnesium, multiply each isotope's mass by the percent abundance and add these products together:

78.7% x 23.99 amu =   18.88 amu

10.1% x 24.99 amu =   2.52 amu

11.2% x 25.98 amu =   <u>2.91 amu</u>
                        24.31 amu

An element has two common isotopes: 80.4% of the atoms have a mass of 11.01 amu and 19.6% of the atoms have a mass of 10.01 amu. In the space below, set up the equations and calculate the atomic mass for this element. Identify the element.

atomic mass =_____   element _____

3.4 According to the **periodic law** (Sec. 3.4), when elements are arranged in order of increasing atomic number, elements with similar properties occur at periodic intervals. The periodic table represents this statement graphically: elements with similar properties are found in the same **group** (Sec. 3.4) or vertical column. The horizontal rows are known as **periods** (Sec. 3.). A steplike line in the periodic table separates the **metals** (Sec. 3.5) on the left from the **nonmetals** (Sec. 3.5) on the right.

Refer to your periodic table for information to complete the table below:

Element	Group	Period	Metal	Nonmetal
Be	IIA	2	X	
Na				
N				
Br				
O				
Sn				
K				

3.5 The **electron configuration** (Sec. 3.6) of an atom is a statement of the number of electrons the atom has in each **electron subshell** (Sec. 3.5). A shorthand system is used to show electron configurations. For each subshell occupied by electrons, write the number and the symbol for the subshell with a superscript indicating the number of electrons in that subshell.

Example: The electron configuration for $_6C$ is $1s^2 2s^2 2p^2$. This shows that the atom has 2 electrons in the $1s$ subshell, 2 in the $2s$ subshell and 2 in the $2p$ subshell.

Write the electron configurations for the elements below. Use Figure 3-10 in your textbook to determine the order in which the **electron orbitals** (Sec. 3.5) are filled.

Element	Electron configuration
neon	
chlorine	
iron	

3.6 We can classify an element by determining the subshell of its **distinguishing electron** (Sec. 3.7), the last electron added in the electron configuration. If the distinguishing electron is added to an s or a p subshell the element is a **representative element** (Sec. 3.8), and if the p subshell is filled, the element is a **noble gas** (Sec. 3.8). If the distinguishing electron is added to a d subshell, the element is a **transition element** (Sec. 3.8); if it is added to an f subshell, the element is an **inner transition element** (Sec. 3.8).

Write the level of the distinguishing electron for each element below, and indicate the element's classification.

Element	Distinguishing electron	Noble gas	Representative element	Transition element
Mg	$3s^2$		X	
Ti				
Ar				
S				

3.7   The **radioactive decay** (Sec. 3.11) of naturally radioactive atoms results in the emission of three common types of radiation: **alpha particles, beta particles** and **gamma rays** (Sec. 3.12). They differ in mass and charge.

Complete the following table summarizing the three types of radiation:

Type of radiation	Mass number	Charge	Symbol
alpha			
beta			
gamma			

3.8   All **radioactive atoms** (Sec. 3.10) do not decay at the same rate; the more unstable the nucleus, the faster it decays. The **half-life** (Sec. 3.11) of a substance, the amount of time for one-half of a given quantity of radioactive substance to decay, is a measure of the stability of the atomic nucleus.

Carbon-14 is a radioisotope with a half-life of 5730 years.

a.   How many milligrams of an 8.00 mg sample of this radioisotope will remain undecayed after 5730 years?

b.   How many milligrams of an 8.00 mg sample of carbon-14 remain after three half-lives have elapsed?

3.9   The three types of naturally occurring radioactive emissions differ in their ability to penetrate matter and, therefore, in their biological effects. Complete the following table summarizing properties of the three types of radiation:

Type of radiation	Speed	Penetration	Biological damage
alpha			
beta			
gamma			

Self-Test

True-false: Indicate whether the following statements are true or false. If the statement is false, give the word or phrase that may be substituted for the underlined portion to make the statement true.

1.   The <u>nucleus</u> is the smallest particle of an atom.
2.   Most of the mass of an atom is located in the <u>nucleus</u>.
3.   The nucleus contains <u>electrons and protons</u>.
4.   For a neutral atom, the number of protons <u>equals</u> the number of electrons.
5.   The atomic number of an element is the number of <u>protons and neutrons</u>.
6.   The mass number is the total number of <u>protons and electrons</u> in the atom.

7. Isotopes of a specific element have different numbers of <u>neutrons</u> in the nuclei of their atoms.

8. Electron orbitals have different shapes: s-orbitals are <u>spherical</u>.

9. The periodic law states that when elements are arranged in order of <u>increasing atomic number</u>, elements with similar properties occur at periodic intervals.

10. In the modern periodic table, the horizontal rows are called <u>groups</u>.

11. <u>Metals</u> are substances which have a high luster and are malleable.

12. Metals are on the <u>left</u> side of the periodic table.

13. Nonmetals are <u>good</u> conductors of electricity.

14. The elements occupying the d-area of the periodic table are called <u>transition elements</u>.

15. <u>Alpha particles</u> are the same type of radiation as X-rays.

16. Beta particles are more penetrating than <u>alpha particles</u>.

17. The half-life of a radioactive isotope is the length of time needed for <u>all</u> of the substance to decay.

18. An example of a <u>diagnostic</u> use of radioisotopes is the use of radiation to kill cancer cells.

Multiple choice:

19. The nucleus of an atom contains these basic particles:

   a. electrons and protons   b. neutrons and electrons   c. protons and neutrons
   d. only neutrons   e. none of these

20. Isotopes of a specific element vary in the following manner:

   a. electron numbers are different   b. neutron numbers are different
   c. proton numbers are different   d. neutron and proton numbers are different
   e. none of these

21. The element, $^{48}_{22}\text{Ti}$, has the following electron configuration:

   a. $1s^22s^22p^63s^23p^63d^{10}4s^2$ b. $1s^22s^22p^63s^23p^63d^4$
   c. $1s^22s^22p^63s^23p^64s^23d^2$ d. $1s^22s^22p^63s^23p^64s^23d^4$
   e. none of these

22. In the isotope, $^{81}_{35}\text{Br}$, how many neutrons are in the nucleus?

   a. 35   b). 46   c. 81   d. 116   e. none of these

23. How many nucleons are located in an atom of cesium, $^{133}_{55}\text{Cs}$?

   a. 55   b. 78   c. 133   d. 188   e. none of these

24. An element has the electron configuration $1s^22s^22p^63s^23p^64s^23d^{10}4p^6$; the element is:

   a. $_{10}\text{Ne}$ b. $_{54}\text{Xe}$ c. $_{18}\text{Ar}$ d. $_{36}\text{Kr}$ e. none of these

25. In the periodic table, the elements on the far left side are classified as:

   a. metals   b. noble gases   c. transition metals
   d. nonmetals   e. none of these

26. In the periodic table, the elements called noble gases are in:

a. Group IA   b. Group IIA   c. Group VIIA
d. Group VA   e. none of these

27. The distinguishing electron for $_{19}K$ would be found in what subshell?

a. $3s$ b. $2p$ c. $3d$ d. $4s$ e. none of these

28. An alpha particle is made up of

a. two protons and four neutrons   b. two protons and two neutrons
c. two neutrons and two electrons   d. two protons and two electrons
e. none of these

Answers to Practice Exercises

3.1

Atomic number	Mass number	Number of protons	Number of neutrons	Number of electrons	Symbol of element
6	12	6	6	6	C
19	39	19	20	19	K
54	131	54	77	54	Xe
29	64	29	35	29	Cu
35	80	35	45	35	Br

3.2

Isotope	A	Z	Protons	Neutrons	Electrons
$_{20}^{40}Ca$	40	20	20	20	20
$_{18}^{40}Ar$	40	18	18	22	18
$_{11}^{23}Na$	23	11	11	12	11
$_{17}^{37}Cl$	37	17	17	20	17
$_{17}^{35}Cl$	35	17	17	18	17

3.3   80.4% x 11.01 amu =   8.85 amu
19.6% x 10.01 amu = 1.96 amu
                              10.81 amu

Atomic mass = 10.81 amu   element: boron

3.4

Element	Group	Period	Metal	Nonmetal
Be	IIA	2	X	
Na	IA	3	X	
N	VA	2		X
Br	VIIA	4		X
O	VIA	2		X
Sn	IVA	5	X	
K	IA	4	X	

3.5

Element	Electron configuration
neon	$1s^2 2s^2 2p^6$
chlorine	$1s^2 2s^2 2p^6 3s^2 3p^5$
iron	$1s^2 2s^2 2p^6 3s^2 3p^6 4s^2 3d^6$

3.6

Element	Distinguishing electron	Noble gas	Representative element	Transition element
Mg	$3s^2$		X	
Ti	$3d^2$			X
Ar	$3p^6$	X		
S	$3p^4$		X	

3.7

Type of radiation	Mass number	Charge	Symbol
alpha	4	+2	$^4_2\alpha$
beta	0	−1	$^0_{-1}\beta$
gamma	0	0	$^0_0\gamma$

3.8   a.   1/2 x 8.00 mg = 4.00 mg remaining

b.   1/2 x 1/2 x 1/2 x 8.00 mg = 1.00 mg remaining

3.9

Type of radiation	Speed	Penetration	Biological damage
alpha	slow	very little, stopped by skin or paper	ingestion – damages internal organs
beta	faster	penetrating, stopped by aluminum foil	skin burns, ingestion damages internal organs
gamma	fastest (speed of light)	very penetrating, stopped by lead/oncrete	causes serious damage to all tissues

Answers to Self-Test

The numbers in parentheses refer to sections in your textbook.

1. F; electron (3.1) **2.** T (3.1) **3.** F; protons and neutrons (3.1) **4.** T (3.3)

5. F; protons (3.3) **6.** F; protons and neutrons (3.3) **7.** T (3.3) **8.** T (3.6) **9.** T (3.4)

10. F; periods (3.4) **11.** T (3.5) **12.** T (3.5) **13.** F; poor (3.5) **14.** T (3.9)

15. F; gamma rays (3.14) **16.** T (3.14) **17.** F; one-half (3.11) **18.** F; therapeutic (3.15)

19. c (3.1) **20.** b (3.3) **21.** c (3.7) **22.** b (3.3) **23.** c (3.3) **24.** d (3.7)

25. a (3.5) **26.** e; Group VIIIA (3.9) **27.** d (3.8) **28.** b (3.12)

Chapter Overview

The electron configuration of the atoms of an element determines the chemical properties of that element. In this chapter you will identify the valence electrons of an atom using its electron configuration and the element's group number in the periodic table. You will draw Lewis structures showing the valence electrons of atoms, and use these structures to show electron transfer in ionic bond formation and sharing of electrons between atoms in covalent bond formation. You will predict the chemical formulas for ionic solids and for covalently bonded molecules, and practice naming both types of compounds.

Electronegativity differences between atoms determine the characteristics of bonds between atoms. You will classify bonds as ionic or covalent, polar or nonpolar. You will use VSEPR theory to predict the three-dimensional shape of molecules and to determine molecular polarity.

Practice Exercises

4.1   The **valence electrons** (Sec. 4.2) of an atom are the electrons in the outermost electron shell. Write the electron configuration and give the number of valence electrons for each atom and the group number each of the following elements:

Element	Electron configuration	Number of valence electrons	Group number
lithium			
beryllium			
boron			
phosphorus			
sulfur			

4.2   **Lewis structures** (Sec. 4.2) are atomic symbols with one dot for each valence electron placed around the element's symbol. Give the number of valence electrons and draw the Lewis structure for each of the elements or groups below (Use the symbol X as a group symbol.)

Group	Valence electrons	Lewis structure
Group IA		
Group IVA		
Group VIIA		

Element	Valence electrons	Lewis structure
sulfur		
bromine		
magnesium		

4.3   According to the **octet rule** (Sec. 4.3), in compound formation, atoms of elements lose, gain or share electrons in such a way that their electron configurations become identical to that of the noble gas nearest them in the periodic table. Atoms on the left side of the periodic table tend to lose electrons and become positively charged **monoatomic ions** (Sec. 4.19) and those on the right side of the periodic table gain electrons to become negatively charged monoatomic ions.

Complete the table below showing ion formation, and identify the noble gas having the same electron configuration as the ion formed.

Element	Group number	Electrons lost/gained	Ion formed	Noble gas
Na				
Br				
S				
Ca				
N				

4.4   **Ionic compounds** (Sec. 4.6) form when electrons are transferred from metal atoms to nonmetal atoms. **Binary ionic compounds** (Sec. 4.9) are named by naming the metal first, followed by the stem of the nonmetal with the ending -*ide*.

In the table below, show the formation of ionic compounds (as in Example 4.3 in your text), and name the ionic compounds that are formed.

Compound unit	Formation of Lewis structure	Compound name
KBr		
CaI_2		
SrS		
Na_2O		

4.5   Since ionic compounds consist of an alternating array of positive and negative charges, the term **formula unit** (Sec. 4,8) is used to refer to the smallest unit of the compound. Formation of ionic compounds requires a charge balance: the same number of electrons must be lost as are gained.

Practice balancing charges by writing chemical formulas for ionic compounds formed from the elements in the table below:

Elements	Ions formed	Formula unit of ionic compound	Name of ionic compound
potassium and chlorine			
beryllium and iodine			
sodium and sulfur			
aluminum and fluorine		AlF_3	
			Calcium oxide

4.6 **Bonding electrons** (Sec. 4.11) are pairs of valence electrons that are shared between atoms in a covalent bond. **Nonbonding electrons** (Sec. 4.11) are pairs of electrons that are not involved in sharing. Molecules tend to be stable when each atom in the molecule has an octet of electrons (bonding and nonbonding).

Draw the Lewis structures for one molecule of each of these covalent compounds. Circle the bonding electrons in each **single covalent bond** (Sec. 4.12) in the molecule:

H⊙Br ̈ HBr	 F₂	 BrI	 H₂O

4.7 In a **double covalent bond** (Sec. 4.12), two atoms share two pairs of electrons; in a **triple covalent bond** (Sec. 4.12), two atoms share three pairs of electrons.

For each molecule below, indicate the number of bonds of each type and the number of nonbonding electron pairs.

Formula	C_2H_4	CS_2	HCN	CH_4O
Lewis structure	H:C::C:H ̈H ̈H	:S::C::S:	H:C:::N:	H H:C:O:H H
single bonds				
double bonds				
triple bonds				
nonbonding electron pairs				

4.8   In Lewis structures for molecules, the shared electron pairs may be represented with lines. Rewrite the structures in exercise 4.7 by replacing the bonding electron pairs with a line to show the covalent bond between atoms. Include the nonbonding electron pairs as dots:

H—C=C—H
   | |
   H H
 	
C_2H_4
	CS₂	HCN	CH₄O

4.9   According to **VSEPR theory** (Sec. 4.14), the geometry of a molecule is determined by the number of electron pairs (both bonding and nonbonding) around the central atom of a molecule. Double and triple bonds each count as a single electron pair. The shape of the molecule is the one in which the electron pairs are furthest from one another.

The shape that will give the greatest distance between electron clouds is: for two electron pairs, 180° bond angle (linear); for three electron pairs, 120° bond angle (trigonal planar or angular); for four electron pairs, 109° bond angle (tetrahedral, trigonal pyramidal, or angular). Use these guidelines to predict the shape of the following molecules (See Chemistry at a Glance 4B in your textbook).

In the table below you are given the Lewis structure for five molecules. For each molecule, classify the electron pairs around the central atom and use this information to predict the geometry of the molecule:

Lewis structure	VSEPR electron pairs around central atom		Molecular geometry
	Bonding pairs	Nonbonding pairs	
:Br: :Br:C:Br: :Br:			
H:C::O: H			
:S::C::S:			
H:S: H			
:Cl:N:Cl: :Cl:			

4.10   **Electronegativity** (Sec. 4.15) is a measure of the relative attraction that an atom has for the shared electrons in a bond. The electronegativity difference between the two bonded atoms is a measure of the polarity of the bond. If the difference is 2.0 or greater, the bond is ionic.

Calculate the electronegativity difference for each element pair, and indicate whether the bond formed between them will be ionic, **nonpolar covalent** (Sec. 4.17), or **polar covalent** (Sec. 4.17) (Electronegativities are in Figure 4.14 of your textbook.)

Pair of elements	Electronegativity difference	Ionic bond	Nonpolar covalent bond	Polar covalent bond
sodium and fluorine				
bromine and bromine				
sulfur and oxygen				
phosphorus and bromine				

4.11   In naming binary molecular compounds, name the element of lower electronegativity first, followed by the stem of the more electronegative nonmetal and the suffix *-ide*. Include prefixes to indicate the number of atoms of each nonmetal. Name the following molecular compounds:

a. CCl_4 _____

b. CS_2 _____

c. NCl_3 _____

4.12   **Polyatomic ions** (Sec. 4.19) are **covalently bonded** (Sec. 4.1) groups of atoms having a charge. In writing the formula for an ionic compound, treat a polyatomic ion as a unit. If more than one of these ions is required for charge balance, enclose the ion in parentheses and put the number of ions outside the parentheses.

Give the formulas for the ionic compounds prepared by combining the following ions. (See Table 4.3 in your textbook for polyatomic ions):

	Bromide	Nitrate	Carbonate	Phosphate
Sodium	NaBr			
Calcium		$Ca(NO_3)_2$		
Ammonium				
Aluminum				

4.13 In naming a compound containing a metal with a variable ionic charge, use a Roman numeral after the metal name to indicate the charge on the metal ion.

Give the formulas and the names of the ionic compounds prepared by combining the ions in the following table. Two have been done for you as examples.

	F^-	N^{3-}	SO_4^{2-}	ClO_2^-
K^+		K_3N potassium nitride		
Pb^{2+}				
Fe^{3+}			$Fe_2(SO_4)_3$ iron (III) sulfate	

Self-Test

True-false: Indicate whether the following statements are true or false. If the statement is false, give the word or phrase that may be substituted for the underlined portion to make the statement true.

1. Lewis structures show the number of <u>inner electrons</u> of an atom.

2. A negative ion is formed when an element <u>loses</u> an electron.

3. Metals tend to <u>gain</u> electrons to attain the configuration of a noble gas.

4. Bromine would accept an electron to attain the configuration of the noble gas <u>krypton</u>.

5. <u>An ionic bond</u> is formed through the sharing of one or more pairs of electrons between atoms.

6. The most stable electron configuration is that of <u>the noble gases</u>.

7. <u>An ionic compound</u> is formed from a metal, which can donate electrons, and a nonmetal, which can accept electrons.

8. In binary ionic compounds, the full name of the metallic element is given <u>first</u>.

9. Covalent bond formation between nonmetal atoms involves electron <u>transfer</u>.

10. <u>Nonbonding</u> electrons are pairs of valence electrons that are not shared between atoms having a covalent bond.

11. A nitrogen molecule, N_2, would have a <u>double</u> covalent bond between the two nitrogen atoms.

12. Carbon can form <u>multiple</u> covalent bonds with other nonmetallic elements.

13. According to VSEPR theory, the electron pairs in the valence shell arrange themselves to <u>maximize</u> the repulsion between the electron pairs.

14. According to VSEPR theory, a water molecule, H_2O, would have <u>a linear</u> arrangement of the valence electron clouds.

15. According to VSEPR theory convention, single and triple bonds are <u>equal</u>.

16. Electronegativity is a measure of the relative <u>repulsion</u> that an atom has for the shared electrons in a bond.

17. Electronegativity values <u>increase</u> from left to right across periods in the Periodic Table.

18. The bond between fluorine and bromine would be <u>a nonpolar covalent</u> bond.

Multiple choice:

19. The binary ionic compound, RbI, would be called;

    a. Rubidium(I) iodide   b. Rubidium iodate   c. Rubidium iodine
    d. Rubidium iodide   e. none of these

20. The formula for the binary ionic compound silver sulfide is;

    a. SiS   b. AgS   c. Ag_2S d. AgS_2 e. none of thes

21. Which representative element would have the same number of valence electrons as calcium?

    a. potassium   b. barium   c. silicon
    d. aluminum   e. none of these

22. The electron configuration of a noble gas would be:

    a. $1s^2 2s^2$ b. $1s^2 2s^2 2p^4$ c. $1s^2 2s^2 2p^6 3s^2 3p^2$
    d. $1s^2 2s^2 2p^6 3s^2 3p^6$ e. none of these

23. The electron configuration for the sulfide ion (S^{2-}) would be:

    a. $1s^2 2s^2 2p^6$ b. $1s^2 2s^2 2p^6 3s^2 3p^4$ c. $1s^2 2s^2 2p^6 3s^2 3p^6$
    d. $1s^2 2s^2 2p^6 3s^2 3p^6 4s^2$ e. none of these

24. In the formula, Na_3N, the total number of electrons accepted by the nitrogen is:

    a. 1   b. 2   c. 3   d. 4   e. none of these

25. The Lewis structure for a Group VA element would have dots representing the following number of electrons:

    a. two   b. three   c. four   d. five   e. none of these

26. Which of the following pairs of elements would form a covalent bond?

    a. sulfur and oxygen   b. potassium and iodine   c. magnesium and bromine
    d. calcium and fluorine   e. none of these

27. Which of the following pairs of elements would form a nonpolar covalent bond?

    a. nitrogen and oxygen   b. fluorine and fluorine   c. calcium and iodine
    d. potassium and bromine   e. none of these

28. Which of the following pairs of elements would form a polar covalent bond?

    a. carbon and carbon   b. bromine and bromine   c. potassium and fluorine
    d. sodium and oxygen   e. none of these

29. How many valence electrons are found in a triple covalent bond?

    a. 2   b. 3   c. 4   d. 6   e. none of these

30. What types of electron pairs are used in VSEPR calculations?

    a. core electrons   b. bonding electrons   c. nonbonding electrons
    d. both b. and c.   e. none of these

31. Which element would be more electronegative than chlorine?

    a. sulfur   b. lithium   c. bromine   d. carbon   e. none of these

32. The correct name for the binary molecular compound, SO_3 is:

    a. sulfur oxide   b. oxygen sulfide   c. sulfur trioxide
    d. trioxygen sulfide   e. none of these

Answers to Practice Exercises

4.1

Element	Electron configuration	Number of valence electrons	Group number
lithium	$1s^2 2s^1$	1	IA
beryllium	$1s^2 2s^2$	2	IIA
boron	$1s^2 2s^2 2p^1$	3	IIIA
phosphorus	$1s^2 2s^2 2p^6 3s^2 3p^3$	5	VA
sulfur	$1s^2 2s^2 2p^6 3s^2 3p^4$	6	VIA

4.2

Group	Valence electrons	Lewis structure
Group IA	1	X
Group IVA	4	·X·
Group VIIA	7	:X:

Element	Valence electrons	Lewis structure
sulfur	6	:S:
bromine	7	·Br:
magnesium	2	·Mg·

4.3

Element	Group number	Electrons lost/gained	Ion formed	Noble gas
Na	IA	1 lost	Na^+	Ne
Br	VIIA	1 gained	Br^-	Kr
S	VIA	2 gained	S^{2-}	Ar
Ca	IIA	2 lost	Ca^{2+}	Ar
N	VA	3 gained	N^{3-}	Ne

4.4

Compound unit	Formation of Lewis structure	Compound name
KBr	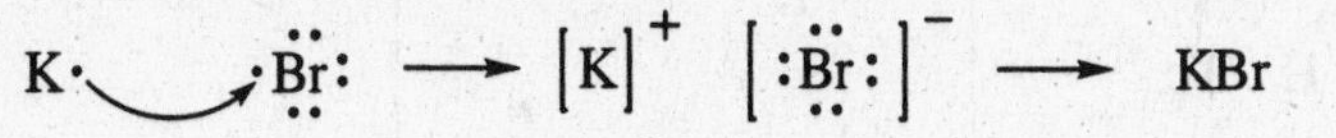 K· ⟶ :Br: ⟶ [K]$^+$ [:Br:]$^-$ ⟶ KBr	potassium bromide
CaI$_2$	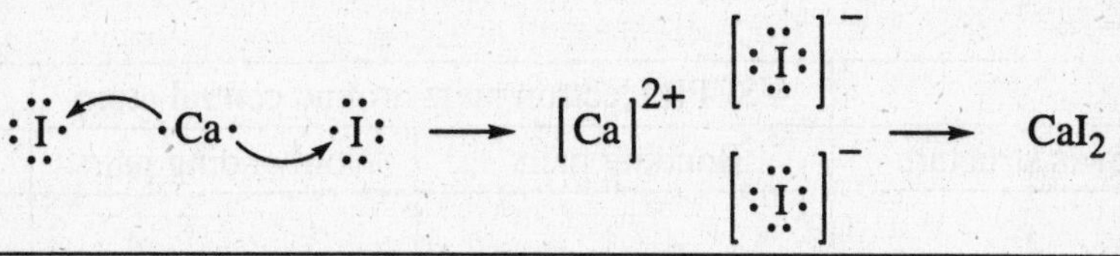 :I· ·Ca· ·I: ⟶ [Ca]$^{2+}$ [:I:]$^-$ [:I:]$^-$ ⟶ CaI$_2$	calcium iodide
SrS	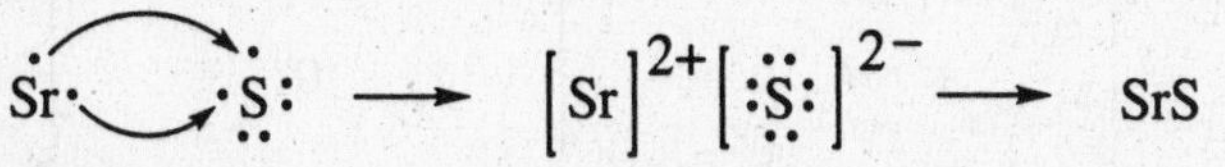 Sr· ·S: ⟶ [Sr]$^{2+}$[:S:]$^{2-}$ ⟶ SrS	strontium sulfide
Na$_2$O	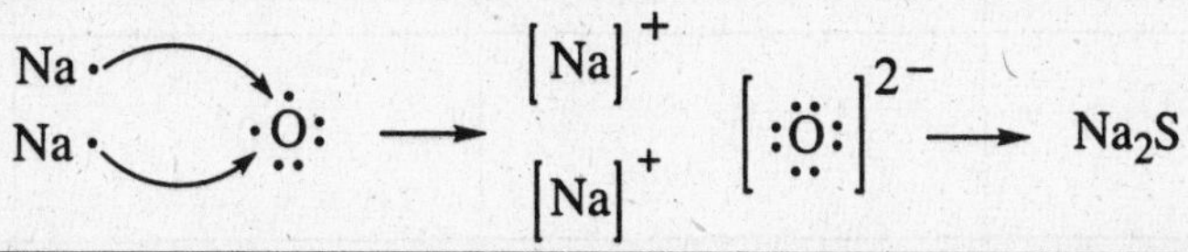 Na· Na· ·O: ⟶ [Na]$^+$ [Na]$^+$ [:O:]$^{2-}$ ⟶ Na$_2$S	sodium oxide

4.5

Elements	Ions formed	Formula unit	Name of ionic compound
potassium and chlorine	K$^+$, Cl$^-$	KCl	potassium chloride
beryllium and iodine	Be^{2+}. I$^-$	BeI$_2$	beryllium iodide
sodium and sulfur	Na$^+$. S^{2-}	Na$_2$S	sodium sulfide
aluminum and fluorine	Al^{3+}, F$^-$	AlF$_3$	aluminum fluoride
calcium and oxygen	Ca^{2+}, O^{2-}	CaO	calcium oxide

4.6

 H⊙Br:	 :F⊙F:	 :Br⊙I:	 H⊙O: ⊙ H

4.7

Lewis structure	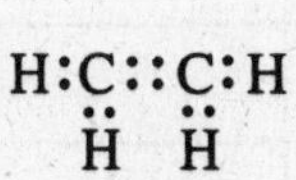 H:C::C:H Ḧ Ḧ	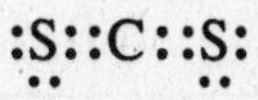 :S::C::S:	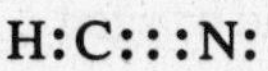 H:C:::N:	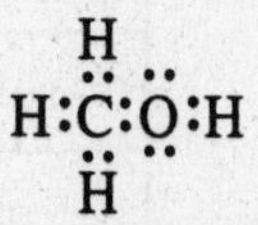 H H:C:O:H Ḧ
single bonds	4	0	1	5
double bonds	1	2	0	0
triple bonds	0	0	1	0
nonbonding pairs	0	4	1	2

4.8

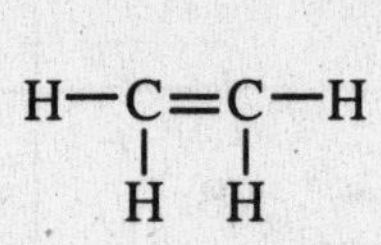 H—C=C—H | | H H	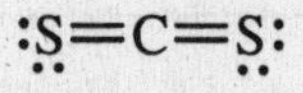 :S̈=C=S̈:	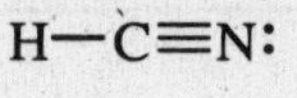 H—C≡N:	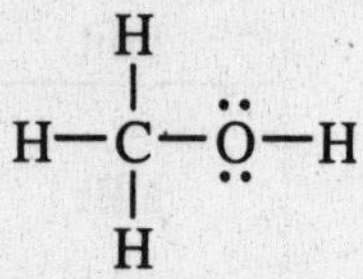 H | H—C—Ö—H | H

4.9

Lewis structure	VSEPR electron pairs around central atom		Molecular geometry
	Bonding pairs	Nonbonding pairs	
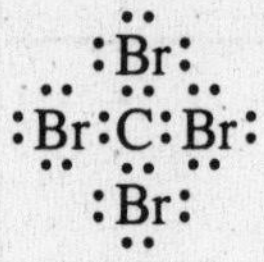 :B̈r: :B̈r:C:B̈r: :B̈r:	4	0	tetrahedron
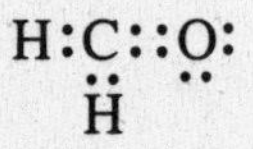 H:C::Ö: Ḧ	3	0	trigonal planar
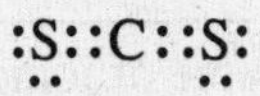 :S̈::C::S̈:	2	0	linear
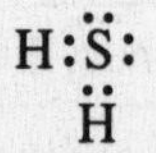 H:S̈: Ḧ	2	2	angular
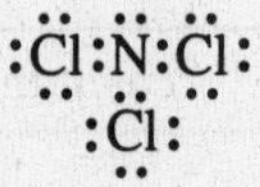 :C̈l:N:C̈l: :C̈l:	3	1	trigonal pyramid

4.10

Pair of elements	Electronegativity difference	Ionic bond	Nonpolar covalent bond	Polar covalent bond
sodium and fluorine	3.1	X		
bromine and bromine	0.0		X	
sulfur and oxygen	1.0			X
phosphorus and bromine	0.7			X

4.11   a. CCl_4 carbon tetrachloride

b. CS_2 carbon disulfide

c. NCl_3 nitrogen trichloride

4.12

	Bromide	Nitrate	Carbonate	Phosphate
Sodium	NaBr	$NaNO_3$	Na_2CO_3	Na_3PO_4
Calcium	$CaBr_2$	$Ca(NO_3)_2$	$CaCO_3$	$Ca_3(PO_4)_2$
Ammonium	NH_4Br	NH_4NO_3	$(NH_4)_2CO_3$	$(NH_4)_3PO_4$
Aluminum	$AlBr_3$	$Al(NO_3)_3$	$Al_2(CO_3)_3$	$AlPO_4$

4.13

	F^-	N^{3-}	SO_4^{2-}	ClO_2^-
K^+	KF potassium fluoride	K_3N potassium nitride	K_2SO_4 potassium sulfate	$KClO_2$ potassium chlorite
Pb^{2+}	PbF_2 lead(II) fluoride	Pb_3N_2 lead(II) nitride	$PbSO_4$ lead(II) sulfate	$Pb(ClO_2)_2$ lead(II) chlorite
Fe^{3+}	FeF_3 iron(III) fluoride	FeN iron(III) nitride	$Fe_2(SO_4)_3$ iron(III) sulfate	$Fe(ClO_2)_3$ iron(III) chlorite

Answers to Self-Test

The numbers in parentheses refer to sections in your textbook.
1. F; outermost or valence electrons (4.2) **2.** F; gains (4.4) **3.** F; lose (4.5) **4.** T (4.5)
5. F; covalent (4.4 and 4.10) **6.** T (4.3) **7.** T (4.6) **8.** T (4.9) **9.** F; sharing (4.10) **10.** T (4.11)
11. F; triple (4.12) **12.** T (4.12) **13.** F; minimize (4.14) **14.** F; an angular (4.14) **15.** T (4.14)
16. F; attraction (4.15) **17.** T (4.15) **18.** F; a polar covalent (4.16) **19.** d (4.9) **20.** c (4.9)
21. b (4.2) **22.** d (4.3) **23.** c (4.3) **24.** c (4.6) **25.** d (4.2) **26.** a (4.10) **27.** b (4.16)
28. e (4.16) **29.** d (4.12) **30.** d (4.14) **31.** e (4.15) **32.** c (4.18)

Chemical Calculations: Formula Masses, Moles, and Chemical Equations   Chapter 5

Chapter Overview

Calculation of the ratios and masses of the substances involved in chemical reactions is very important in many chemical processes. Central to these calculations is the concept of the mole, a convenient counting unit for atoms and molecules.

In this chapter you will learn to determine the formula mass of substances and the number of moles of substances. You will practice writing and balancing chemical equations, and you will learn to use these equations in determining amounts of substances that react and are produced in chemical reactions.

Practice Exercises

5.1   The **formula mass** (Sec. 5.1) of a compound is the sum of the atomic masses of the atoms in its formula. Calculate the formula mass for these compounds:

Formula	Calculations with atomic masses	Formula mass
KBr		
$CaCl_2$		
Na_2CO_3		
$(NH_4)_3PO_4$		

5.2   The **mole** (Sec. 5.2) is a useful unit for counting numbers of atoms and molecules. The number of particles in a mole is 6.02×10^{23}, known as **Avogadro's number** (Sec. 6.2). Set up and complete the calculations below:

Moles	Relationship with units	Number of atoms
1.00 mole of helium atoms		
2.60 moles of sodium atoms		
0.316 mole of argon atoms		

5.3   The mass of one mole, called **molar mass** (Sec. 5.3), of any substance is its formula mass expressed in grams.

Using dimensional analysis and the correct conversion factor (either moles/gram or grams/mole), find either the mass or the number of moles for the substances below:

Given quantity	Relationship with units	Calculated
1.00 mole H_2O		g H_2O
2.53 moles H_2O		g H_2O
0.519 mole H_2O		g H_2O
1.00 g NaBr		mole NaBr
417 g NaBr		mole NaBr
0.322 g NaBr		mole NaBr

5.4   The subscripts in a chemical formula show the number of atoms of each element per formula unit. They also show the number of moles of atoms of each element in one mole of the substance (atoms/molecule = moles of atoms/moles of molecules). Determine the moles of carbon atoms in the following problems:

Moles of compound	Carbon atoms/molecule	Moles of carbon atoms
1.00 mole of $C_{10}H_{16}$ (limonene)		
13.5 moles of C_2H_6O (ethanol)		
0.705 mole of C_5H_{12} (pentane)		

5.5   The number of grams of one substance cannot be compared directly to the number of grams of another substance; however, moles can be related to moles quite easily by looking at subscripts in chemical formulas. Figure 5.7 in your textbook gives you a map of the steps to take in solving problems involving grams and moles.

a.   How many moles of oxygen atoms are contained in the following molar amounts of NO_2? (Hint: What is the relationship between moles of NO_2 and moles of oxygen atoms in the formula NO_2?)

moles of NO_2	Relationship with units	Moles of oxygen atoms
2.00 moles NO_2		moles O atoms
0.500 mole NO_2		moles O atoms

b. How many grams of oxygen atoms can be obtained from the following molar amounts of NO_2? (Hint: What is the relationship between moles of oxygen atoms and grams of oxygen atoms?)

Moles of NO_2	Relationship with units	Mass of oxygen atoms
3.00 moles NO_2		g of O atoms
0.400 mole NO_2		g of O atoms

c. What is the mass of oxygen atoms contained in the following masses of NO_2? (Hint: What is the relationship between the mass of NO_2 and the number of moles of NO_2?)

Mass of NO_2	Relationship with units	Mass of oxygen atoms
3.00 g of NO_2		g of O atoms
0.500 g of NO_2		g of O atoms

5.6 The description of a chemical reaction can be expressed efficiently with the formulas and symbols of a **chemical equation** (Sec. 5.6). The substances that react (the reactants) are placed on the left, and those that are produced (products) are on the right. The arrow in the chemical equation is read as "to produce"; plus signs on the left side mean "reacts with," and plus signs on the right are read as "and."

Write chemical equations for the following chemical reactions:

a. Hydrogen chloride reacts with sodium hydroxide to produce sodium chloride and water.

b. Silver nitrate and potassium bromide react with one another to produce silver bromide and potassium nitrate.

5.7 To be most useful, chemical equations must be **balanced** (Sec. 5.6); that is, the number of atoms of each element must be the same on each side of the equation. A suggested method for balancing equations is in Section 5.6 of your textbook. Remember: use the **coefficients** (Sec. 5.6) to balance equations, but do not change the subscripts within the formulas. Balance the following equations:

a. $H_2 + Cl_2 \rightarrow HCl$

b. $AgNO_3 + H_2S \rightarrow Ag_2S + HNO_3$

c. $P + O_2 \rightarrow P_2O_3$

d. $HCl + Ba(OH)_2 \rightarrow BaCl_2 + H_2O$

5.8   Balanced chemical equations are useful in telling us what amounts of products we can expect from a given amount of reactant, or how much reactant to use for a specific amount of product. The chemical equation tells us the ratios of the numbers of atoms and molecules involved in the reaction, and it also tells us the ratios of the numbers of moles of the substances involved.

The diagram in Figure 5.9 of your textbook will help you to determine the sequence of steps to use in solving the following types of problems.

Balanced equation: $4Na + O_2 \rightarrow 2Na_2O$

a.   How many atoms of Na are required to produce four molecules of Na_2O?

b.   How many moles of Na_2O could be produced from 0.300 mole of Na?

c.   How many moles of Na_2O could be produced from 2.18 g of sodium?

d.   How many grams of Na_2O could be produced from 5.15 g of Na?

5.9   Balance the equation: $Al + Cl_2 \rightarrow AlCl_3$

a.   How many moles of chlorine gas will react with 0.160 mole of aluminum?

b.   How many grams of aluminum chloride could be produced from 5.27 moles of aluminum?

c.   What is the maximum number of grams of aluminum chloride that could be produced from 14.0 g of chlorine gas?

d.   How many grams of chlorine gas would be needed to react with 0.746 g of aluminum?

Self-Test

True-false: Indicate whether the following statements are true or false. If the statement is false, give the word or phrase that may be substituted for the underlined portion to make the statement true.

1. The mass of one mole of helium atoms would be <u>the same as</u> the mass of one mole of gold atoms.

2. In a balanced equation, the total number of atoms on the reactants side is <u>equal to</u> the total number of atoms on the products side.

3. Formula masses are calculated on the ${}^{16}_{8}O$ <u>relative-mass</u> scale.

4. In balancing a chemical equation, do not change the <u>coefficients</u> within the formulas.

5. Atomic mass and formula mass are both expressed in <u>amu</u>.

6. Carbon dioxide is a colorless, odorless gas that <u>does not</u> support combustion.

7. The number of <u>atoms</u> in a mole of H_2O is equal to 6.02×10^{23}.

8. One mole of glucose ($C_6H_{12}O_6$) contains <u>6 moles</u> of carbon atoms.

9. In a chemical equation, the products are the materials that are <u>consumed</u>.

10. A mole of nitrogen gas contains 6.02×10^{23} nitrogen <u>atoms</u>.

Multiple choice:

11. How many atoms are contained in 6.8 moles of calcium?

    a. 4.1×10^{24} b. 1.6×10^{26} c. 6.2×10^{22}
    d. 3.3×10^{23} e. none of these

12. What is the formula mass for iron(III) carbonate [$Fe_2(CO_3)_3$]?

    a. 115.86 g/mole   b. 287.57 g/mole   c. 291.73 g/mole
    d. 171.71 g/mole   e. none of these

13. How many grams are contained in 4.72 moles of $NaHCO_3$?

    a. 84.1   b. 283   c. 264   d. 396   e. none of these

14. A mole of butane contains four moles of carbon atoms and ten moles of hydrogen atoms. Its formula is:

    a. C_6H_6 b. H_6C_{10} c. C_4H_{10} d. H_4C_6 e. none of these

15. What is the total number of moles of all atoms in 1.00 mole of $(NH_4)_2CO_3$?

    a. 21.0   b. 14.0   c. 9.00   d. 8.00   e. none of these

16. The conversion factor used in changing grams of O_2 to moles of O_2 is:

    a. 16.00 g/1 mole   b. 1 mole/16.00 g   c. 32.00 mole/1 g
    d. 1 mole/32.00 g   e. none of these

17. When oxygen gas and hydrogen gas combine to form water, which of the following is true? (Hint: Write the balanced equation.)

    a. 2 moles of O_2 produce 1 mole of H_2O b. 2 moles of H_2 react with 1 mole of H_2O
    c. 1 mole of O_2 produces 1 mole of H_2O d. 2 moles of H_2 produce 2 moles of H_2O
    e. none of these

18. Using the equation you wrote in question 17, find the mass of water in grams that would be produced by the complete reaction of 4.00 g of oxygen gas.

    a. 4.50 g   b. 36.0 g   c. 7.32 g

    d. 14.7 g   e. none of these

19. The mass of 0.560 mole of methanol, CH_4O, would equal:

    a. 32.0 amu   b. 32.0 g   c. 17.9 amu

    d. 17.9 g   e. none of these

20. 42.0 g of ethanol, C_2H_6O, would equal:

    a. 1.10 moles   b. 0.911 mole   c. 1.10 amu

    d. 0.911 amu   e. none of these

21. After balancing the equation below, calculate the sum of all the coefficients in the balanced equation:

$$C_2H_6 + O_2 \rightarrow CO_2 + H_2O$$

    a. 4   b. 9   c. 15   d. 19   e. none of these

Use the following balanced equation to answer questions 22 and 23 below.

$$CH_4 + 2O_2 \rightarrow CO_2 + 2H_2O$$

22. How many moles of water would be produced from 0.420 mole of methane (CH_4)?

    a. 2.00 moles   b. 0.210 mole   c. 0.420 mole

    d. 0.840 mole   e. none of these

23. How many grams of methane would be used to produce 10.0 g of water?

    a. 2.22 g   b. 4.44 g   c. 8.88 g

    d. 10.0 g   e. none of these

Answers to Practice Exercises

5.1

Formula	Calculations with atomic masses	Formula mass
KBr	39.10 + 79.90	119.00 amu
$CaCl_2$	40.08 + 2(35.45)	110.98 amu
Na_2CO_3	2(22.99) + 12.01 + 3(16.00)	105.99 amu
$(NH_4)_3PO_4$	3[14.01 + 4(1.01)] + 30.97 + 4(16.00)	149.12 amu

5.2

Moles	Relationship with units	Number of atoms
1.00 mole of helium atoms	$1.00 \text{ mole} \times \dfrac{6.02 \times 10^{23} \text{ atoms}}{1 \text{ mole}} =$	6.02×10^{23}
2.60 moles of sodium atoms	$2.60 \text{ moles} \times \dfrac{6.02 \times 10^{23} \text{ atoms}}{1 \text{ mole}} =$	1.57×10^{24}
0.316 mole of argon atoms	$0.316 \text{ mole} \times \dfrac{6.02 \times 10^{23} \text{ atoms}}{1 \text{ mole}} =$	1.90×10^{23}

5.3

Given quantity	Relationship with units	Calculated
1.00 mole H_2O	$1.00 \text{ mole } H_2O \times \dfrac{18.0 \text{ g } H_2O}{1.00 \text{ mole } H_2O} =$	18.0 g H_2O
2.53 moles H_2O	$2.53 \text{ mole } H_2O \times \dfrac{18.0 \text{ g } H_2O}{1.00 \text{ mole } H_2O} =$	45.5 g H_2O
0.519 mole H_2O	$0.519 \text{ mole } H_2O \times \dfrac{18.0 \text{ g } H_2O}{1.00 \text{ mole } H_2O} =$	9.34 g H_2O
1.00 g NaBr	$1.00 \text{ g NaBr} \times \dfrac{1.00 \text{ mole NaBr}}{103 \text{ g NaBr}} =$	0.00971 mole NaBr
417 g NaBr	$417 \text{ g NaBr} \times \dfrac{1.00 \text{ mole NaBr}}{103 \text{ g NaBr}} =$	4.05 mole NaBr
0.322 g NaBr	$0.322 \text{ g NaBr} \times \dfrac{1.00 \text{ mole NaBr}}{103 \text{ g NaBr}} =$	0.00313 mole NaBr

5.4

Moles of compound	Carbon atoms/molecule	Moles of carbon atoms
1.00 mole of $C_{10}H_{16}$ (limonene)	10	10(1.00) = 10.0 moles
13.5 moles of C_2H_6O (ethanol)	2	2(13.5) = 27.0 moles
0.705 mole of C_5H_{12} (pentane)	5	5(0.705) = 3.53 moles

5.5   a.

moles of NO_2	Relationship with units	Moles of oxygen atoms
2.00 moles NO_2	2.00 moles NO_2 x $\dfrac{2 \text{ moles O}}{1 \text{ mole } NO_2}$ =	4.00 moles of O atoms
0.500 mole NO_2	0.500 moles NO_2 x $\dfrac{2 \text{ moles O}}{1 \text{ mole } NO_2}$ =	1.00 moles of O atoms

b.

Moles of NO_2	Relationship with units	Mass of oxygen atoms
3.00 moles NO_2	3.00 moles NO_2 x $\dfrac{2 \text{ moles O}}{1 \text{ mole } NO_2}$ x $\dfrac{16.0 \text{ g O}}{1 \text{ mole O}}$ =	96.0 g of O atoms
0.400 mole NO_2	0.400 moles NO_2 x $\dfrac{2 \text{ moles O}}{1 \text{ mole } NO_2}$ x $\dfrac{16.0 \text{ g O}}{1 \text{ mole O}}$ =	12.8 g of O atoms

c.

Mass of NO_2	Relationship with units	Mass of oxygen atoms
3.00 g of NO_2	3.00 g NO_2 x $\dfrac{1.00 \text{ mole } NO_2}{46.0 \text{ g } NO_2}$ x $\dfrac{2 \text{ moles O}}{1 \text{ mole } NO_2}$ x $\dfrac{16.0 \text{ g O}}{1 \text{ mole O}}$ =	2.09 g of O atoms
0.500 g of NO_2	0.500 g NO_2 x $\dfrac{1.00 \text{ mole } NO_2}{46.0 \text{ g } NO_2}$ x $\dfrac{2 \text{ moles O}}{1 \text{ mole } NO_2}$ x $\dfrac{16.0 \text{ g O}}{1 \text{ mole O}}$ =	0.348 g of O atoms

5.6   a. $HCl + NaOH \rightarrow NaCl + H_2O$

   b. $AgNO_3 + KBr \rightarrow AgBr + KNO_3$

5.7   a. $H_2 + Cl_2 \rightarrow 2HCl$

b. $2\,AgNO_3 + H_2S \rightarrow Ag_2S + 2\,HNO_3$

c. $4\,P + 3\,O_2 \rightarrow 2\,P_2O_3$

d. $2\,HCl + Ba(OH)_2 \rightarrow BaCl_2 + 2H_2O$

5.8   Balanced equation: $4Na + O_2 \rightarrow 2Na_2O$

a. $4 \text{ molecules } Na_2O \times \dfrac{2 \text{ atoms Na}}{1 \text{ molecule } Na_2O} = 8 \text{ atoms Na}$

b. $0.300 \text{ mole Na} \times \dfrac{2 \text{ moles } Na_2O}{4 \text{ moles Na}} = 0.150 \text{ mole } Na_2O$

c. $2.18 \text{ g Na} \times \dfrac{1 \text{ mole Na}}{22.99 \text{ g Na}} \times \dfrac{2 \text{ moles } Na_2O}{4 \text{ moles Na}} = 0.0474 \text{ mole } Na_2O$

d. $5.15 \text{ g Na} \times \dfrac{1 \text{ mole Na}}{22.99 \text{ g Na}} \times \dfrac{2 \text{ moles } Na_2O}{4 \text{ moles Na}} \times \dfrac{61.98 \text{ g } Na_2O}{1 \text{ mole } Na_2O} = 6.94 \text{ g } Na_2O$

5.9 $2Al + 3Cl_2 \rightarrow 2AlCl_3$ (molar mass of $AlCl_3$: 133.33 g/mole)

a. $0.160 \text{ mole Al} \times \dfrac{3 \text{ moles } Cl_2}{2 \text{ moles Al}} = 0.240 \text{ mole } Cl_2$

b. $5.27 \text{ moles Al} \times \dfrac{2 \text{ moles } AlCl_3}{2 \text{ moles Al}} \times \dfrac{133.33 \text{ g } AlCl_3}{1 \text{ mole } AlCl_3} = 703 \text{ g } AlCl_3$

c. $14.0 \text{ g } Cl_2 \times \dfrac{1 \text{ mole } Cl_2}{70.90 \text{ g}} \times \dfrac{2 \text{ moles } AlCl_3}{3 \text{ moles } Cl_2} \times \dfrac{133.33 \text{ g } AlCl_3}{1 \text{ mole } AlCl_3} = 17.6 \text{ g } AlCl_3$

d. $0.746 \text{ g Al} \times \dfrac{1 \text{ mole Al}}{26.98 \text{ g Al}} \times \dfrac{3 \text{ moles } Cl_2}{2 \text{ moles Al}} \times \dfrac{70.90 \text{ g } Cl_2}{1 \text{ mole } Cl_2} = 2.94 \text{ g } Cl_2$

Answer to Self-Test

1. F; different than or less than (5.3) 2. T (5.6) 3. F; $^{12}_{6}C$ relative mass (5.1)
4. F; subscripts (5.6) 5. T (5.1) 6. T (5.3) 7. F; molecules (5.4) 8. T (5.4)
9. F; produced (5.6) 10. F; molecules (5.4) 11. a (5.4) 12. c (5.1) 13. d (5.4)
14. c (5.4) 15. b (5.4) 16. d (5.5) 17. d (5.6) 18. a (5.8) 19. d (5.3) 20. b (5.3)
21. d (5.6) 22. d (5.8) 23. b (5.8)

Chapter Overview

The physical states of matter and the behavior of matter in these states are determined by the behavior of the particles (atoms, molecules, ions) of which matter is made. The movements and interactions of these particles are described by the kinetic molecular theory of matter.

In this chapter you will study the five statements of the kinetic molecular theory and the ways that these statements explain the physical behavior of matter. You will use the gas laws to describe quantitatively various changes in the conditions of pressure, temperature, and volume of matter in the gaseous state. You will study three types of **intermolecular forces** (Sec. 6.13) that affect liquids and solids and their changes of state.

Practice Exercises

6.1 According to the **kinetic molecular theory of matter** (Sec. 6.1), the differing physical properties of the **solid, liquid, and gaseous states** of matter (Sec. 6.2) are determined by the **potential energy** (cohesive forces) and the **kinetic energy** (disruptive forces) of that state.

In the table below indicate which form of energy is dominant in a given state and write in a word or two to describe how this dominance affects the physical properties below:

State	Cohesive forces	Disruptive forces	Density	Compressibility	Motion of particles
Gas					
Liquid					
Solid					

6.2 Gases can be described by simple quantitative relationships called **gas laws** (6.3). According to **Boyle's law** (6.4), the volume of a gas is inversely proportional to the **pressure** (Sec. 6.3) applied to it if the temperature is held constant: $P_1 \times V_1 = P_2 \times V_2$

a. Complete the following table using Boyle's law. Rearrange the equation to solve for the missing variable.

P_1	V_1	P_2	V_2
6.0 atm	8.0 L	2.0 atm	? L
3.0 atm	14 L	? atm	7.0 L
6.0 atm	? L	4.0 atm	3.0 L
? atm	4.0 L	2.0 atm	8.0 L

b. The pressure on 2.2 L of helium is changed from 2.8 atm to 5.6 atm. What is the new volume?

6.3   According to **Charles's law** (Sec. 6.5), the volume of a gas at constant pressure is proportional to the Kelvin temperature of the gas: $V_1/T_1 = V_2/T_2$

a.   Complete the following table using Charles's law. Remember that temperatures must be in Kelvin degrees. Rearrange the equation to solve for the needed variable:

V_1	T_1	V_2	T_2
6.00 L	127°C	3.00 L	? K
3.00 L	227°C	? L	327°C
3.00 L	? K	6.00 L	127°C
? L	127°C	5.00 L	427°C

b.   The temperature of 5.00 L of gas is reduced from 327°C to 127°C. What is the new volume of the gas?

6.4   The gas laws can be combined into a single equation called the **combined gas law** (Sec. 6.6):

$$\frac{P_1 \times V_1}{T_1} = \frac{P_2 \times V_2}{T_2}$$

This equation can be used to solve for one of the variables if the others are known.

Example: A sample of helium gas occupies 425 mL at 127°C and 745 mm Hg. If the gas is compressed to 375 mL and the temperature rises to 227°C, what is the new pressure of the gas?

$$P_2 = \frac{P_1 \times V_1 \times T_2}{T_1 \times V_2} = \frac{745 \text{ mm Hg} \times 425 \text{ mL} \times 500 \text{ K}}{400 \text{ K} \times 375 \text{ mL}} = 1060 \text{ mm Hg}$$

a.   The volume of a gas is 5.72 L at 30°C and 1.25 atm. If the gas is heated to 50°C and compressed to a volume of 4.50 L, what will be the new pressure?

b.   A gas at 514°C and 338 mm Hg was cooled to 311°C and 507 mm Hg. What would be the final volume of the gas, if the initial volume was 14.2 L?

6.5   Combining the three gas laws gives an equation that describes the state of a gas at a single set of conditions. This equation, $PV = nRT$, is called the **ideal gas equation** (Sec. 6.7). T is measured on the Kelvin scale, and the value of R (the ideal gas constant) is 0.0821 atm·L/mole·K.

Example: Calculate the pressure, in atmospheres, of 2.00 moles of nitrogen gas in a 6.00 L flask at a temperature of 127°C.

First, solve the ideal gas equation for P. Then substitute the known quantities for the variables:

$$P = \frac{n \times R \times T}{V} = \frac{2.00 \text{ moles} \times 0.0821 \frac{\text{atm L}}{\text{mole K}} \times 400 \text{ K}}{6.00 \text{ L}} = 10.9 \text{ atm}$$

Rearrange the ideal gas equation to solve for the correct variable in completing the following combined gas law problems.

a.   What is the volume of 1.49 moles of helium with a pressure of 1.21 atm at 224°C?

b.   What would be the temperature of neon gas, if 0.339 mole of neon gas is in a 5.72 liter tank and the pressure gauge reads 2.53 atm?

6.6   **Dalton's law of partial pressures** (Sec. 6.8) states that the total pressure exerted by a mixture of gases is the sum of the **partial pressures** (Sec. 6.8) of the individual gases: $P_T = P_1 + P_2 + P_3 + \ldots$ Using Dalton's law of partial pressures, complete the following problems:

a.   What is the total pressure exerted by a mixture of helium and argon if the partial pressures of helium and argon are: $P_{He} = 270$ mm Hg and $P_{Ar} = 400$ mm Hg?

b.   In a mixture of O_2, CO_2 and CO, the CO_2 has a partial pressure of 341 mm Hg and the CO a pressure of 114 mm Hg. If the total pressure of the mixture is 744 mm Hg, what is the partial pressure of oxygen gas?

6.7   Use the following table summarizing the gas laws to review your knowledge of this chapter:

Law	Quantities held constant	Variables	Equation
Boyle's Law			
			$\dfrac{V_1}{T_1} = \dfrac{V_2}{T_2}$
Combined Gas Law			
Ideal Gas Law	R = 0.0821 L atm/mole K		
Dalton's Law of Partial Pressures			

6.8   A **change of state** (Sec. 6.9) is a process in which matter changes from one state to another. If heat is absorbed, the change is **endothermic** (Sec. 6.9); if heat is released during the process, it is **exothermic** (Sec. 6.9).

a.   Complete the following table with the correct term for the physical change involved.

	To solid	To liquid	To gas
From solid	XXXXXXX		
From liquid		XXXXXXX	
From gas			XXXXXXX

b.   In the table above, indicate for each physical change whether it is endothermic (A) or exothermic (B).

6.9   An **intermolecular force** (Sec. 6.13) is an attractive force that acts between a molecule and another molecule. **Dipole-dipole interactions** occur between polar molecules. **Hydrogen bonds** are strong dipole-dipole interactions that occur when hydrogen is covalently bonded to a very small, electronegative atom (F, O, or N). **London forces** are weak forces that occur when there is a temporary uneven distribution of electrons in a molecule.

In the table below, indicate which type of intermolecular force would be the dominant attractive force between molecules in the following substances:

Molecule	Hydrogen bonds	Dipole-dipole interaction	London forces
IBr			
H_2O			
HF			
NH_3			
Cl_2			
CO			

Self-Test

True-false: Indicate whether the following statements are true or false. If the statement is false, give the word or phrase that may be substituted for the underlined portion to make the statement true.

1. Boyle's law states for a given mass of gas at constant temperature, the volume of a gas varies directly with pressure.

2. Charles's law states for a given mass of gas at constant pressure, the volume of a gas is directly proportional to the temperature.

3. When a fixed volume of gas is cooled, its pressure increases.

4. If the temperature and number of moles of gas remain constant, then doubling the volume of a gas will double the pressure of the gas..

5. One atmosphere is the pressure required to support 760 mm of Hg.

6. The total pressure exerted by a mixture of gases is equal to the sum of the partial pressures.

7. The vapor pressure of a liquid decreases as temperature increases.

8. As temperature increases, the velocity of molecules in a liquid decreases.

9. The energy resulting from the attractions and repulsions between particles in matter is a part of that matter's potential energy.

10. Liquids are very compressible because there is a lot of empty space between particles.

11. A balloon filled with helium and left to warm in the sun will decrease in volume.

12. A volatile liquid is one that has a high vapor pressure.

13. An equilibrium state may exist between a liquid and a gas in a closed container.

Multiple choice:

14. Oxygen gas with a volume of 5.00 L at 1 atm and 273 K was heated to 402 K and the pressure was doubled. What was the new volume of the oxygen gas?

   a. 3.68 L   b. 5.00 L   c. 1.71 L   d. 14.7 L   e. none of these

15. Two gases, nitrogen and oxygen, in the same container, have a total pressure of 600 mm Hg. If the partial pressure of oxygen equals the partial pressure of nitrogen, what is the partial pressure of oxygen?

    a. 600 mm Hg   b. 400 mm Hg   c. 300 mm Hg
    d. 200 mm Hg   e. none of these

16. How many moles of helium are in a 28.4 L balloon at 45°C and 1.03 atm?

    a. 0.271 mole   b. 0.0453 mole   c. 6.98 moles
    d. 1.12 moles   e. none of these

17. The strongest intermolecular forces between water molecules are:

    a. ionic bonds   b. covalent bonds   c. hydrogen bonds
    d. London forces   e. none of these

18. Hydrogen bonding would not occur between two molecules of which of these compounds?

    a. HF   b. CH_4 c. CH_3NH_2 d. H_2O e. CH_3OH

19. London forces would be the strongest attractive forces between two molecules of which of these substances?

    a. HF   b. BrCl   c. F_2 d. H_2O e. none of these

20. Liquids that have significant hydrogen bonding, compared to liquids that have no hydrogen bonding, would have:

    a. higher vapor pressure   b. higher boiling point
    c. lower temperature of condensation   d. greater tendency to evaporate
    e. none of these

21. The pressure on 526 mL of gas is increased from 755 mm Hg to 974 mm Hg. If temperature remains constant, what is the new volume?

    a. 408 mL   b. 633 mL   c. 215 mL   d. 387 mL   e. none of these

22. Which of the following changes is endothermic?

    a. condensation   b. freezing   c. sublimation
    d. deposition   e. none of these

23. What will increase the pressure of a gas in a closed container?

    a. decreasing the temperature of the gas
    b. adding more gas to the container
    c. increasing the volume of the container
    d. replacing the gas with the same number of moles of a different gas
    e. both b and c

Answers to Practice Exercises

6.1

State	Cohesive forces	Disruptive forces	Density	Compressibility	Motion of particles
Gas		X	low	large	random, fast-moving
Liquid	X	X	high	small	slide freely, but do not separate
Solid	X		high	small	vibration in a fixed position

6.2 a.

P_1	V_1	P_2	V_2
6.0 atm	8.0 L	2.0 atm	24 L
3.0 atm	14 L	6.0 atm	7.0 L
6.0 atm	2.0 L	4.0 atm	3.0 L
4.0 atm	4.0 L	2.0 atm	8.0 L

b. $V_2 = \dfrac{P_1 \times V_1}{P_2} = \dfrac{2.8 \text{ atm} \times 2.0 \text{ L}}{5.6 \text{ atm}} = 1.0 \text{ L}$

6.3 a.

V_1	T_1	V_2	T_2
6.00 L	127°C	3.00 L	200 K
3.00 L	227°C	3.60 L	327°C
3.00 L	200 K	6.00 L	127°C
2.86 L	127°C	5.00 L	427°C

b. $V_2 = \dfrac{V_1 \times T_2}{T_1} = \dfrac{5.00 \text{ L} \times 400 \text{ K}}{600 \text{ K}} = 3.33 \text{ L}$

6.4 a. $P_2 = \dfrac{P_1 \times V_1 \times T_2}{T_1 \times V_2} = \dfrac{1.25 \text{ atm} \times 5.72 \text{ L} \times 323 \text{ K}}{4.50 \text{ L} \times 303 \text{ K}} = 1.69 \text{ atm}$

b. $V_2 = \dfrac{P_1 \times V_1 \times T_2}{T_1 \times P_2} = \dfrac{338 \text{ mm Hg} \times 14.2 \text{ L} \times 584 \text{ K}}{787 \text{ K} \times 507 \text{ mm Hg}} = 7.02 \text{ L}$

6.5 a. $V = \dfrac{n \times R \times T}{P} = \dfrac{1.49 \text{ moles} \times 0.0821 \text{ atm L/mole K} \times 497 \text{ K}}{1.21 \text{ atm}} = 50.2 \text{ L}$

b. $T = \dfrac{P \times V}{n \times K} = \dfrac{2.53 \text{ atm} \times 5.72 \text{ L}}{0.339 \text{ mole} \times 0.0821 \text{ atm L/mole K}} = 5.20 \times 10^2 \text{ K}$

6.6 a. $P_{\text{total}} = P_{\text{He}} + P_{\text{Ar}} = 270 \text{ mm Hg} + 400 \text{ mm Hg} = 670 \text{ mm Hg}$

b. $P_{\text{total}} = P_{O_2} + P_{CO_2} + P_{CO}$

$P_{O_2} = P_{\text{total}} - P_{CO_2} - P_{CO} = 744 \text{ mm Hg} - 114 \text{ mm Hg} - 341 \text{ mm Hg} = 289 \text{ mm Hg}$

6.7

Law	Quantities held constant	Variables	Equation
Boyle's Law	temperature, number of moles of gas	pressure, volume	$P_1V_1 = P_2V_2$
Charles's Law	pressure , number of moles of gas	volume, temperature	$\dfrac{V_1}{T_1} = \dfrac{V_2}{T_2}$
Combined Gas Law	number of moles of gas	pressure, volume, temperature	$\dfrac{P_1 \times V_1}{T_1} = \dfrac{P_2 \times V_2}{T_2}$
Ideal Gas Law	$R = 0.0821$ L atm/mole K	pressure, volume, temperature, number of moles of gas	$PV = nRT$
Dalton's Law of Partial Pressures	volume, temperature, number of moles of gas	partial pressures, total pressure	$P_T = P_1 + P_2 + P_3 + \ldots$

6.8

	To solid	To liquid	To gas
From solid	XXXXX	melting A	sublimation A
From liquid	freezing B	XXXXX	evaporation A
From gas	deposition B	condensation B	XXXXX

6.9

Molecule	Hydrogen bonds	Dipole-dipole interaction	London forces
IBr		X	
H$_2$O	X		
HF	X		
NH$_3$	X		
Cl$_2$			X
CO		X	

Answers to Self-Test

The numbers in parentheses refer to sections in your textbook:
1. F; varies inversely (6.4) **2.** T (6.5) **3.** F; decreases (6.4) **4.** F; halve (6.4)
5. T (6.3) **6.** T (6.8) **7.** F; increases (6.11) **8.** F; increases (6.1) **9.** T (6.1)
10. F; gases (6.2) **11.** F; increase (6.5) **12.** T (6.11) **13.** T (6.11) **14.** a (6.6) **15.** c (6.8)
16. d (6.7) **17.** c (6.13) **18.** b (6.13) **19.** c (6.13) **20.** b (6.13) **21.** a (6.4) **22.** c (6.9)
23. b (6.7)

Chapter Overview

Many chemical reactions take place in solutions, particularly in water solutions. The properties of water make it a vital part of all living systems.

In this chapter you will define terms associated with solutions, study how solutions form, and calculate the concentrations of solutions using various units. You will study osmotic pressure and the factors that control the important process of osmosis.

Practice Exercises

7.1 A **solution** (Sec. 7.1) is a homogeneous mixture consisting of a **solvent** (Sec. 7.1) and one or more **solutes** (Sec. 7.1). The solvent is the substance present in the greatest amount.

In the table below, identify the solute and the solvent in each of the solutions:

Solution	Solute	Solvent
10.0 g of potassium chloride in 70.0 g of water		
80.0 g of ethyl alcohol in 50.0 g of water		
40.0 g of potassium iodide in 55.0 g of water		
30.0 mL of ethyl alcohol in 40.0 mL of methyl alcohol		

7.2 The **solubility** (Sec. 7.2) of a substance depends on many factors: the temperature of the solution, the pressure on the solution, the nature of the solvent. The rate at which a substance dissolves to form a solution depends on how fast the particles come in contact with the solvent.

In the table below indicate whether each of the changes in conditions would affect the solubility or the rate of solution of a given solute.

Change of conditions	Solubility in water	Rate of solution in water
raising the temperature		
crushing or grinding the solute		
adding more solute		
agitating the solute and solvent		

7.3 The solubility of a molecular substance in water depends on the polarity of the solute. In general, a polar solute will be soluble in a polar solvent (such as water). Many ionic solids are soluble in water, but not all; several factors influence the solubility of ionic compounds.

Complete the table below, giving the nature of each solute and using it to predict the solubility of the substance in water. (For ionic solids, see Table 7.2 in your textbook.)

Formula unit	Ionic, polar or nonpolar	Very soluble	Slightly soluble
NaCl			
BaSO$_4$			
Cl$_2$			
CCl$_4$			
HCl			
KNO$_3$			

7.4 The **concentration** (Sec. 7.5) of a solution is the amount of solute present in a specified amount of solution. One way of expressing concentration is **percent by mass** (Sec. 7.5), the mass of solute divided by the mass of solution multiplied by 100.

$$\text{percent by mass} = \%(m/m) = \frac{\text{mass of solute}}{\text{mass of solution}} \times 100$$

a. What is the percent by mass, %(m/m), concentration of NaCl in a solution prepared by dissolving 14.8 g of NaCl in 122 g of water?

b. How many grams of NaCl were added to 225 g of water to prepare a 7.52%(m/m) NaCl solution?

7.5 **Percent by volume** (Sec. 7.5) is a percentage unit used when the solute and the solvent are both liquids or both gases.

$$\text{percent by volume} = \%(v/v) = \frac{\text{volume of solute}}{\text{volume of solution}} \times 100$$

a. Calculate the percent by volume for the following solution:
25.0 mL of ethyl alcohol is added to enough water to make 155 mL of solution.

b. If 165 mL of ethyl alcohol are added to enough water to give a volume of 425 mL, what is the percent by volume of the resulting solution?

7.6   Another commonly used concentration unit is **mass-volume percent** (Sec. 7.5), the mass of
            solute divided by the volume of solution:

$$\text{mass-volume percent} = \%(m/v) = \frac{\text{mass of solute (g)}}{\text{volume of solution (mL)}} \times 100$$

a. How many grams of KCl must be added to 250.0 mL of water to prepare a 9.82%(m/v)
    solution?

b. Calculate the mass-volume percent of the following solution:
    10.5 g of sugar added to enough water to make a solution having a volume of 164 mL.

7.7   The **molarity** (Sec. 7.5) of a solution is a ratio giving the number of moles of solute per liter
            of solution:

$$\text{Molarity(M)} = \frac{\text{moles of solute}}{\text{liters of solution}}$$

a. What is the molarity of a solution that contains 1.44 moles of NaCl in 2.50 L of
    solution?

b. A solution with a volume of 425 mL is prepared by dissolving 2.64 moles of $CaCl_2$ in
    water. What is the molarity?

c. If 7.21 g of KCl is dissolved in water to prepare 0.333 L of solution, what is the
    molarity?

7.8   The molarity of a solution can be used as a conversion factor to relate liters of solution to
            moles of solute. Use dimensional analysis to solve for the correct variable in the following
            problems.

a. How many moles of $CaCl_2$ were dissolved in 3.55 L of a 1.47 M solution?

b. How many grams of $CaCl_2$ were used to prepare 0.250 L of a 0.143 M $CaCl_2$ solution?

7.9 **Dilution** (Sec. 7.6) is the process in which more solvent is added to a solution in order to lower its concentration. The simple relationship used for dilution is: the concentration of the stock solution times the volume of the stock solution is equal to the concentration of the diluted solution times the volume of the diluted solution.

$$C_s \times V_s = C_d \times V_d$$

a. If 255 mL of water is added to 325 mL of a 0.477 M solution, what is the molarity of the new solution?

b. How many milliliters of water would have to be added to 250.0 mL of a 2.33 M solution to prepare a 0.551 M solution?

7.10 **Colligative properties** (Sec. 7.7) of solutions are those physical properties affected by the concentration of a solute. In the following table, tell whether a given change in a solution's condition will cause the value of each colligative property to increase or decrease.

Change in conditions	Boiling point	Freezing point	Vapor pressure
adding NaCl to an aqueous solution			
adding water to an aqueous sugar solution			
putting antifreeze in the radiator of a car			

7.11 **Osmosis** (Sec. 7.8) is the movement of water across a **semipermeable membrane** (Sec. 7.8) from a more dilute solution to a more concentrated solution. **Osmotic pressure** (Sec. 7.8), the amount of pressure necessary to stop this flow of water, depends on the number of particles of solute in the solutions.

Osmolarity (Sec. 7.8) is the product of the molarity of the solution and the number of particles produced per formula unit when the solute dissociates:

Osmolarity = molarity x i

Complete the following table on osmolarity.

Molarity of solutions	Osmolarity
3 M KCl	
2 M KCl + 1 M glucose	
3 M $CaBr_2$ + 2 M glucose	
2 M $CaBr_2$ + 1 M KBr	

7.12 Water flows across the semipermeable membrane of a cell from a solution of higher solute concentration to one of lower solute concentration. A solution outside the cell is classified with reference to the solution within the cell: a **hypotonic** solution has a lower concentration than the concentration of the solution within the cell, a **hypertonic** solution has a higher concentration of solute, and an **isotonic** solution has the same solute concentration (Sec. 7.8).

Indicate which way water will flow across the cell membranes of red blood cells under each of these conditions:

Condition	Water flows into cells	Water flows out of cells
Cells immersed in concentrated NaCl solution		
Solution around cells is hypotonic		
Cells immersed in an isotonic solution		
Hypertonic solution surrounds cells		
Cells immersed in pure water		
Cells immersed in physiological saline solution (0.9 % m/v)		

Self-Test

True-false: Indicate whether the following statements are true or false. If the statement is false, give the word or phrase that may be substituted for the underlined portion to make the statement true.

1. A saturated solution contains the maximum amount of <u>solute</u> that will dissolve in the solution.

2. Colligative properties are properties that depend on the <u>amount of solute</u> dissolved in a given mass of solution.

3. Water flows through an osmotic semipermeable membrane toward the <u>more dilute</u> solution..

4. Hypertonic solutions contain a <u>smaller</u> number of solute particles than the intracellular fluid.

5. In a salt water solution, the <u>solute</u> is water.

6. Increasing the pressure on a solution in which the solute is a gas will <u>increase</u> the solubility of the gas.

7. Undissolved solute is in equilibrium with dissolved solute in a <u>saturated solution</u>.

8. An unsaturated solution is <u>always</u> a dilute solution.

9. Carbon dioxide is <u>more</u> soluble in water when pressure increases.

10. A polar gas, such as NO_2, is <u>insoluble</u> in water.

11. Percent by mass (% m/m) is mass of solute divided by mass of <u>solvent</u>, x 100.

12. Red blood cells in a hypotonic solution may undergo <u>hemolysis</u>.

13. <u>The same number of moles</u> of NaCl are in 225 mL of a 1.55 M NaCl solution as are in 450 mL of a 1.55 M NaCl solution.

14. Dialysis can be used to remove <u>dissolved ions</u> from solutions containing large molecules.

Multiple choice:

15. Which of the following compounds would have a very low solubility in water?

   a. NaCl
   b. CCl_4
   c. $CaCl_2$
   d. HCl
   e. all would dissolve

16. A solution containing 10.0 g of NaCl (formula weight 58.5) in 0.500 L of solution would have what molarity?

   a. 0.342 M
   b. 0.200 M
   c. 0.500 M
   d. 0.174 M
   e. none of these

17. How many milliliters of 5.0 M potassium bromide solution would be needed to prepare 1.00 L of 3.0 M potassium bromide solution?

   a. 170 mL
   b. 300 mL
   c. 600 mL
   d. 750 mL
   e. none of these

18. How much solute is present in 200 mL of a 0.50 M solution of HCl in water?

   a. 1.0 moles
   b. 0.050 mole
   c. 0.25 mole
   d. 0.10 mole
   e. none of these

19. If 25 mL of a 3.0 M NaCl solution is diluted to give a solution whose molarity is 0.15 M, what is the volume of the new solution?

   a. 0.75 L
   b. 0.50 L
   c. 250 mL
   d. 600 mL
   e. none of these

20. The osmolarity of a solution that is 2 M $CaCl_2$ and 2 M glucose is:

   a. 4 M
   b. 6 M
   c. 8 M
   d. 10 M
   e. none of these

21. Which of the following solutions would be isotonic with 0.1 M NaCl?

   a. 0.5 M $CaCl_2$
   b. 0.2 M glucose
   c. 0.1 M sucrose
   d. 0.05 $Ca(NO_3)_2$
   e. none of these

22. Water flows out of red blood cells placed in which of the following solutions?

   a. hypotonic
   b. isotonic
   c. hypertonic
   d. both a and c
   e. none of these

23. What mass-volume percent %(m/v) would result from dissolving 5.00 g of NaCl in enough water to form 50.0 mL of saline solution?

   a. 5.00%(m/v)
   b. 9.09%(m/v)
   c. 10.0%(m/v)
   d. 20.0%(m/v)
   e. none of these

24. Dissolving 15 g of NaCl in 85 g of water would yield a solution that is what percent by mass, %(m/m)?

   a. 5.7%(m/m)
   b. 18%(m/m)
   c. 15%(m/m)
   d. 21%(m/m)
   e. none of these

Answers to Practice Exercises

7.1

Solution	Solute	Solvent
10.0 g of potassium chloride in 70.0 g of water	10.0 g potassium chloride	70.0 g of water
80.0 g of ethyl alcohol in 50.0 g of water	50.0 g of water	80.0 g of ethyl alcohol
40.0 g of potassium iodide in 55.0 g of water	40.0 g of potassium iodide	55.0 g of water
30.0 mL of ethyl alcohol in 40.0 mL of methyl alcohol	30.0 mL of ethyl alcohol	40.0 mL of methyl alcohol

7.2

Change of conditions	Solubility in water	Rate of solution in water
raising the temperature	depends on solute (may increase or decrease)	increases
crushing or grinding the solute	no effect	increases
adding more solute	no effect	increases
agitating the solute and solvent	no effect	increases

7.3

Formula unit	Ionic, polar or nonpolar	Very soluble	Slightly soluble
NaCl	ionic	X	
$BaSO_4$	ionic		X
Cl_2	nonpolar		X
CCl_4	nonpolar		X
HCl	polar	X	
KNO_3	ionic	X	

7.4   a. $\%(m/m) = \dfrac{\text{mass of solute}}{\text{mass of solution}} \times 100 = \dfrac{14.8 \text{ g NaCl}}{14.8 \text{ g NaCl} + 122 \text{ g H}_2\text{O}} \times 100 = 10.8\%(m/m)$

   b. 100 g solution − 7.52 g NaCl = 92.48 g H_2O

   $225 \text{ g H}_2\text{O} \times \dfrac{7.52 \text{ g NaCl}}{92.48 \text{ g H}_2\text{O}} = 18.3 \text{ g NaCl}$

7.5   a. $\%(v/v) = \dfrac{\text{volume of solute}}{\text{volume of solution}} \times 100 = \dfrac{25.0 \text{ mL of solute}}{155 \text{ mL of solution}} \times 100 = 16.1\%$

   b. $\%(v/v) = \dfrac{\text{volume of solute}}{\text{volume of solution}} \times 100 = \dfrac{165 \text{ mL of solute}}{425 \text{ mL of solution}} \times 100 = 38.8\%$

7.6   a. $\%(m/v) = 9.82\% = \dfrac{9.82 \text{ g KCl}}{100 \text{ mL solution}} \times 100$

mass of KCl = %(m / v) x volume of solution

mass of KCl = $\dfrac{9.82 \text{ g KCl}}{100 \text{ mL solution}} \times 250 \text{ mL of solution} = 24.6 \text{ g KCl}$

b. $\%(m/v) = \dfrac{\text{mass of solute (g)}}{\text{volume of solution (mL)}} \times 100 = \dfrac{10.5 \text{ g solute}}{164 \text{ mL solution}} \times 100 = 6.40\%$

7.7   a. $\text{Molarity}(M) = \dfrac{\text{moles of solute}}{\text{liters of solution}} = \dfrac{1.44 \text{ moles NaCl}}{2.50 \text{ L}} = 0.576 \text{ M}$

b. $425 \text{ mL} \times \dfrac{1 \text{ L}}{1000 \text{ mL}} = 0.425 \text{ L}$

$\text{Molarity}(M) = \dfrac{\text{moles of solute}}{\text{liters of solution}} = \dfrac{2.64 \text{ moles CaCl}_2}{0.425 \text{ L}} = 6.21 \text{ M}$

c. $7.21 \text{ g} \times \dfrac{1.00 \text{ mole}}{74.6 \text{ g}} = 9.66 \times 10^{-2} \text{ moles KCl}$

$M = \dfrac{\text{moles of solute}}{\text{liters of solution}} = \dfrac{9.66 \times 10^{-2} \text{ moles}}{0.333 \text{ L}} = 0.290 \text{ M}$

7.8   a. $\dfrac{1.47 \text{ moles}}{1 \text{ L}} \times 3.55 \text{ L} = 5.22 \text{ moles CaCl}_2$

b. $\dfrac{0.143 \text{ mole}}{1 \text{ L}} \times \dfrac{111 \text{ g}}{1 \text{ mole}} \times 0.250 \text{ L} = 3.97 \text{ g CaCl}_2$

7.9   a. $(C_s \times V_s = C_d \times V_d)$

$C_d = \dfrac{C_s \times V_s}{V_d} = \dfrac{0.477 \text{ M} \times 325 \text{ mL}}{325 \text{ mL} + 255 \text{ mL}} = 0.267 \text{ M}$

b. $V_d = \dfrac{C_s \times V_s}{C_d} = \dfrac{2.33 \text{ M} \times 250.0 \text{ mL}}{0.551 \text{ M}} = 1057 \text{ mL}$

$V_d - V_s = 1057 \text{ mL} - 250.0 \text{ mL} = 807 \text{ mL water added}$

7.10

Change in conditions	Boiling point	Freezing point	Vapor pressure
adding NaCl to an aqueous solution	increases	decreases	decreases
adding water to an aqueous sugar solution	decreases	increases	increases
putting antifreeze in the radiator of a car	increases	decreases	decreases

7.11

Molarity of solutions	Osmolarity
3 M KCl	6
2 M KCl + 1 M glucose	5
3 M CaBr$_2$ + 2 M glucose	11
2 M CaBr$_2$ + 1 M KBr	8

7.12

Condition	Water flows into cells	Water flows out of cells
Cells immersed in concentrated NaCl solution		X
Solution around cells is hypotonic	X	
Cells immersed in an isotonic solution	no flow in or out	
Hypertonic solution surrounds cells		X
Cells immersed in pure water	X	
Cells immersed in physiological saline solution (0.9 % m/v)	no flow in or out	

Answers to Self-Test

The numbers in parentheses refer to sections in your textbook:
1. T (7.2) **2**. F; number of particles (7.7) **3**. F; more concentrated (7.8)
4. F; larger (7.8) **5**. F; solvent (7.1) **6**. T (7.2) **7**. T (7.2)
8. F; sometimes (7.2) **9**. T (7.2) **10**. F; soluble (7.4) **11**. F; solution (7.5) **12**. T (7.8)
13. F; fewer moles (7.5) **14**. T (7.9) **15**. b (7.4) **16**. a (7.5) **17**. c (7.6) **18**. d (7.5)
19. b (7.6) **20**. c (7.8) **21**. b (7.8) **22**. c (7.8) **23**. c (7.5) **24**. c (7.5)

Chemical Reactions

Chapter 8

Chapter Overview

Chemical reactions are the means by which new substances are formed. The concepts of collision theory explain how and under what conditions reactions take place.

In this chapter you will learn to recognize four basic types of chemical reactions. You will identify oxidizing agents and reducing agents in redox reactions. You will study factors that affect the rate of a chemical reaction. Not all chemical reactions go to completion; you will learn which conditions determine the position of an equilibrium state.

Practice Exercises

8.1 In a **chemical reaction** (Sec. 8.1) at least one new substance is produced as the result of chemical change. Most chemical reactions can be classified in five categories. Classify the following reactions as **combination, decomposition, single replacement, double replacement,** or **combustion reactions** (Sec. 8.1):

Reaction	Classification
a. $2\,NaNO_3 \rightarrow 2\,NaNO_2 + O_2$	
b. $H_2 + Cl_2 \rightarrow 2\,HCl$	
c. $2\,C_2H_6 + 7O_2 \rightarrow 4CO_2 + 6H_2O$	
d. $AgNO_3 + KBr \rightarrow AgBr + KNO_3$	
e. $Cu + 2\,AgNO_3 \rightarrow 2\,Ag + Cu(NO_3)_2$	

8.2 **Oxidation-reduction reactions** (also known as **redox reactions**) (Sec. 8.2) are defined in three different ways: 1) Reactions in which reactants gain or lose oxygen atoms. 2) Reactions in which reactants gain or lose hydrogen atoms. 3) Reactions in which reactants gain or lose electrons. Complete the table below to check your understanding of these definitions:

Definition	Oxidation	Reduction
Oxygen-based	gain of oxygen atoms	
Hydrogen-based		
Electron-based		

The most useful of these three definitions of redox reactions is _____

because _____

8.3 If ions are present in the reactants and products in an equation, the electron-based definition is useful in determining whether a given reaction is a redox reaction. The oxygen-based and hydrogen-based definitions are useful when the reactants and products are molecular.

Determine whether the following reactions are redox or nonredox reactions and tell why you made that determination.

Reaction	Redox or nonredox	Reason
$4Al + 3O_2 \rightarrow 2Al_2O_3$		
$Cu + 2AgNO_3 \rightarrow 2Ag + Cu(NO_3)_2$		
$2CuO \rightarrow 2Cu + O_2$		
$AgNO_3 + NaI \rightarrow AgI + NaNO_3$		
$C_2H_2 + 2H_2 \rightarrow C_2H_6$		

8.4   In redox reactions an **oxidizing agent** (Sec. 8.3) accepts electrons and is reduced. A **reducing agent** (Sec. 8.3) loses electrons and is oxidized. For the equations below, identify the oxidizing and reducing agents and the substances oxidized and reduced.

Equation	Substance oxidized	Substance reduced	Oxidizing agent	Reducing agent
$Ca + S \rightarrow CaS$				
$Cl_2 + 2NaI \rightarrow 2NaCl + I_2$				
$2C_2H_6 + 7O_2 \rightarrow 4CO_2 + 6H_2O$				
$4Na + O_2 \rightarrow 2Na_2O$				
$CuO + H_2 \rightarrow Cu + H_2O$				

8.5   According to **collision theory** (Sec. 8.4), a chemical reaction takes place when two reactant particles collide with a certain minimum amount of energy, called **activation energy** (Sec. 8.4), and the proper orientation. In an energy diagram, the activation energy is the energy difference between the energy of the reactants and the top of the energy "hill". Some of this energy is regained during the reaction; in an **exothermic reaction** (Sec. 8.5) energy is given off in product formation, but in an **endothermic reaction** (Sec. 8.5), the energy is absorbed, so that the products are at a higher energy level than the reactants.

Sketch two energy diagrams below and label these parts on each diagram: a. average energy of reactants, b. average energy of products, c. energy absorbed or given off during the reaction, and d. activation energy.

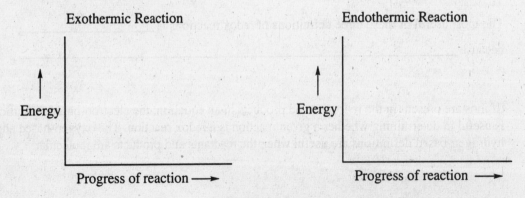

8.6   The **rate of a chemical reaction** (Sec. 8.6) is the rate at which reactants are consumed or products produced in a given time period. Various factors affect the rate of a reaction: the physical nature of reactants, reactant concentrations, reaction temperature, the presence of a **catalyst** (Sec. 8.6).

Indicate whether the conditions listed in the table below would increase or decrease the rate of the following reaction. Explain your answer in terms of collision theory.

$$A(solid) + B \rightarrow C + D + heat$$

Condition	Rate of reaction	Explanation
Decreasing the concentration of B		
Decreasing the temperature of the reaction		
Introduction of an effective catalyst for this reaction		
Increasing the surface area of A by dividing the solid into smaller particles		

8.7   A **reversible reaction** (Sec. 8.7) is a chemical reaction in which two reactions (the forward reaction and the reverse reaction) occur simultaneously. When these two opposing chemical reactions occur at the same rate, the system is said to be at **chemical equilibrium** (Sec. 8.7). Since the rates of the forward and reverse reactions are the same, the concentrations of the reactants and the products remain constant.

$$2P + 3I_2 \rightleftharpoons 2PI_3$$

Le Chatelier's Principle (Sec. 8.8) considers the effects of outside forces on systems at chemical equilibrium. According to this principle, a stress applied to the system can favor the reaction that will reduce the stress, either the forward reaction, in which case more product is formed, or the reverse reaction, in which case more reactants form. Some of the stresses that can cause this readjustment of the equilibrium are: concentration changes, temperature changes, pressure changes.

a.   Indicate what effect each of the conditions below would have on the following exothermic reaction at equilibrium:

$$CH_4(g) + 2O_2(g) \rightleftharpoons CO_2(g) + 2H_2O(g) + heat$$

Condition	Change in equilibrium	Explanation
Increasing the concentration of O_2		
Increasing the temperature of the reaction		
Introduction of an effective catalyst for this reaction		
Increasing the pressure exerted on the reaction		

b. Indicate what effect each of the conditions below would have on the following endothermic reaction

$$N_2(g) + 2O_2(g) + heat \rightleftharpoons 2NO_2(g)$$

Condition	Change in equilibrium	Explanation
Increasing the concentration of O_2		
Increasing the temperature of the reaction		
Introduction of an effective catalyst for this reaction		
Increasing the pressure exerted on the reaction		

Self-Test

True-false: Indicate whether the following statements are true or false. If the statement is false, give the word or phrase that may be substituted for the underlined portion to make the statement true.

1. The reaction $2CuO \rightarrow 2Cu + O_2$ is an example of a <u>single-replacement</u> reaction.

2. Many air pollutants are the result of <u>combustion reactions</u>.

3. In an <u>endothermic</u> reaction, more energy is stored in the product molecule bonds than was stored in the reactant molecule bonds.

4. A substance that is <u>oxidized</u> loses electrons.

5. A reducing agent <u>gains</u> electrons.

6. Adding heat to an <u>endothermic</u> reaction helps the reaction to go toward the products side.

7. The addition of a catalyst <u>will not change</u> the equilibrium position of a reaction.

8. The rate of a reaction is <u>not affected</u> by the addition of a catalyst.

9. Increasing the concentration of products in an equilibrium reaction shifts the equilibrium toward the <u>product side</u> of the reaction.

10. In an energy diagram for an exothermic reaction, the height of the "hill" corresponds to the <u>energy of the reactants</u>.

Multiple choice:

11. The equation $X + YZ \rightarrow Y + XZ$ is a general equation for which type of reaction?

a. combination   b. single-displacement   c. double-displacement
d. decomposition   e. combustion

12. The reaction of CH_4 with oxygen to produce CO_2 and water is an example of which type of reaction?

   a. combination   b. single-displacement   c. double-displacement
   d. decomposition   e. combustion

13. In the reaction $Zn + Cu(NO_3)_2 \rightarrow Zn(NO_3)_2 + Cu$ the oxidizing agent is:

   a. Zn   b. $Cu(NO_3)_2$ c. $Zn(NO_3)_2$
   d. Cu   e. none of these

14. In the reaction $2Fe_2O_3 + 3C \rightarrow 4Fe + 3CO_2$ the reducing agent is:

   a. Fe_2O_3 b. C   c. Fe
   d. CO_2 e. none of these

15. Which of these factors does **not** affect the rate of a reaction?

   a. the frequency of the collisions   b. the energy of the collisions
   c. the orientation of the collisions   d. the product of the collisions
   e. all of these affect rate

16. For a reaction at equilibrium, the concentrations of the products:

   a. increase rapidly   b. increase slowly
   c. remain the same   d. decrease slowly
   e. none of these

Answer questions 17 through 19 using the general equilibrium equation:

$$A + B \rightleftharpoons C + D + heat$$

17. The rate of the forward reaction could be increased by:

   a. decreasing the concentration of B   b. increasing the concentration of A
   c. increasing the concentration of C   d. both b and c
   e. none of these

18. The position of equilibrium would be shifted to the left by:

   a. adding more B
   b. increasing the temperature
   c. decreasing the temperature
   d. increasing the surface area of A by grinding it
   e. adding a catalyst

19. If more A is added to the reaction mixture at equilibrium:

   a. the amount of C will increase   b. the amount of B will increase
   c. the amount of B will decrease   d. both a and c
   e. none of these

20. In the reaction $2Mg + O_2 \rightarrow 2MgO$ the magnesium is:

   a. reduced and is the oxidizing agent   b. reduced and is the reducing agen
   c. oxidized and is the oxidizing agent   d. oxidized and is the reducing agent
   e. none of these

Answers to Practice Exercises

8.1

Reaction	Classification
a. $2NaNO_3 \rightarrow 2NaNO_2 + O_2$	decomposition
b. $H_2 + Cl_2 \rightarrow 2HCl$	combination
c. $2C_2H_6 + 7O_2 \rightarrow 4CO_2 + 6H_2O$	combustion
d. $AgNO_3 + KBr \rightarrow AgBr + KNO_3$	double-replacement
e. $Cu + 2AgNO_3 \rightarrow 2Ag + Cu(NO_3)_2$	single-replacement

8.2

Definition	Oxidation	Reduction
Oxygen-based	gain of oxygen atoms	loss of oxygen atoms
Hydrogen-based	loss of hydrogen atoms	gain of hydrogen atoms
Electron-based	loss of electrons	gain of electrons

The most useful of these three definitions of redox reactions is <u>the electron-based definition</u> because <u>it is true for all redox reactions, not just those involving gain or loss of oxygen or hydrogen.</u>

8.3

Reaction	Redox or nonredox	Reason
$4Al + 3O_2 \rightarrow 2Al_2O_3$	redox	Al loses electrons and gains oxygen; oxygen gains electrons
$Cu + 2AgNO_3 \rightarrow 2Ag + Cu(NO_3)_2$	redox	Cu loses electrons, Ag gains electrons
$2CuO \rightarrow 2Cu + O_2$	redox	Cu gains electrons, loses oxygen; oxygen gains electrons
$AgNO_3 + NaI \rightarrow AgI + NaNO_3$	nonredox	no loss or gain of electrons
$C_2H_2 + 2H_2 \rightarrow C_2H_6$	redox	C_2H_2 gains hydrogen, H_2 loses electrons

8.4

Equation	Substance oxidized	Substance reduced	Oxidizing agent	Reducing agent
$Ca + S \rightarrow CaS$	Ca	S	S	Ca
$Cl_2 + 2NaI \rightarrow 2NaCl + I_2$	I^-	Cl_2	Cl_2	I^-
$2C_2H_6 + 7O_2 \rightarrow 4CO_2 + 6H_2O$	C_2H_6	O_2	O_2	C_2H_6
$4Na + O_2 \rightarrow 2Na_2O$	Na	O_2	O_2	Na
$CuO + H_2 \rightarrow Cu + H_2O$	H_2	CuO	CuO	H_2

8.5   a. average energy of reactants, b. average energy of products, c. energy absorbed or given off during the reaction, and d. activation energy.

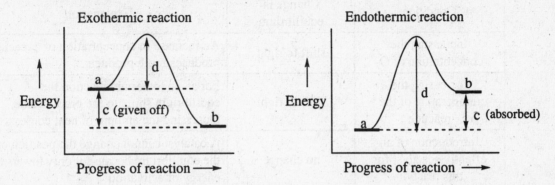

8.6   A(solid) + B → C + D + heat

Condition	Rate of reaction	Explanation
Decreasing the concentration of B	decrease	Fewer molecules collide, so fewer molecules react.
Decreasing the temperature of the reaction	decrease	Lower kinetic energy, lower collision energy, so fewer collisions are effective.
Introduction of an effective catalyst for this reaction	increase	Catalysts provide alternative reaction pathways that have lower energies of activation.
Increasing the surface area of A by dividing the solid into smaller particles	increase	Larger surface area of solid provides more chances for collision.

8.7   a. $CH_4(g) + 2O_2(g) \rightleftharpoons CO_2(g) + 2H_2O(g) + heat$

Condition	Change in equilibrium	Explanation
Increasing the concentration of O_2	shift to right	An increase in concentration of a reactant produces more products.
Increasing the temperature of the reaction	shift to left	The equilibrium shifts to decrease the amount of heat produced.
Introduction of an effective catalyst for this reaction	no change	A catalyst cannot change the position of the equilibrium because it only lowers the energy of activation.
Increasing the pressure exerted on the reaction	no change	Since the moles of gas on each side of the equation are the same, pressure changes have no effect

b. $N_2(g) + 2O_2(g) + heat \rightleftharpoons 2NO_2(g)$

Condition	Change in equilibrium	Explanation
Increasing the concentration of O_2	shift to right	An increase in concentration of a reactant produces more products.
Increasing the temperature of the reaction	shift to right	For an endothermic reaction the equilibrium shifts to the product side, increasing the amount of heat consumed.
Introduction of an effective catalyst for this reaction	no change	A catalyst cannot change the position of the equilibrium because it only lowers the energy of activation.
Increasing the pressure exerted on the reaction	shift to right	An increase in the forward reaction would relieve pressure since there are fewer moles of gas on the right.

Answers to Self-Test

The numbers in parentheses refer to sections in your textbook:
1. F; decomposition (8.1) 2. T (8.1) 3. T (8.5) 4. T (8.2) 5. F; loses (8.3)
6. T (8.5) 7. T (8.7) 8. F; increased (8.6) 9. F; reactant side (8.7)
10. F; activation energy (8.5) 11. b (8.1) 12. e (8.1) 13. b (8.3) 14. b (8.3)
15. d (8.6) 16. c (8.7) 17. b (8.6) 18. b (8.7) 19. d (8.7) 20. d (8.3)

Acids, Bases, and Salts   Chapter 9

Chapter Overview

Acids, bases, and salts play a central role in much of the chemistry that affects our daily lives. Learning the terms and concepts associated with these compounds will give you a greater understanding of the chemistry of the human body, and of the ways in which chemicals are manufactured.

In this chapter you will learn to identify acids and bases according to the Arrhenius and Brønsted-Lowry definitions, write equations for acid and base dissociations in water, and calculate pH, a measure of acidity. You will learn to identify buffers and study the actions of buffers in acidic and basic solutions.

Practice Exercises

9.1   According to the Arrhenius acid-base theory, the **dissociation** (Sec. 9.1) of an **Arrhenius acid** (Sec. 9.1) in water produces hydrogen ions (H^+) and the dissociation of an **Arrhenius base** in water produces hydroxide ions (OH^-). Arrhenius acids and bases have certain properties that help to identify them. In the table below, specify whether each property is that of an acid or of a base.

Property	Acid	Base
has a sour taste		
turns blue litmus red		
has a slippery feel		
turns red litmus blue		
has a bitter taste		

9.2   According to the **Brønsted-Lowry** (Sec. 9.2) definitions, an acid is a proton donor and a base is a proton acceptor. The **conjugate base** (Sec. 9.2) of an acid is the species that remains when an acid loses a proton. The **conjugate acid** (Sec. 9.2) of a base is the species formed when a base accepts a proton.

$$\text{HA} + \text{B} \rightleftharpoons \text{HB}^+ + \text{A}^-$$

HA		B		HB$^+$		A$^-$
Acid		Base		Conjugate acid		Conjugate base

Give the formula of the conjugate acid or base for the following substances:

Base	Conjugate acid
NH_3	
BrO_3^-	
HCO_3^-	

Acid	Conjugate base
HCO_3^-	
$HClO_2$	
HNO_3	

71

9.3   Identify the acid and base in the reactants in the following equations. Identify the conjugate
       acid and conjugate base in the products.

   a. $HClO_3(aq) + H_2O(l) \rightarrow ClO_3^-(aq) + H_3O^+(aq)$

   b. $HNO_2(aq) + OH^-(aq) \rightarrow NO_2^-(aq) + H_2O(l)$

9.4   A **monoprotic acid** (Sec. 9.3) transfers one H^+ ion per molecule during an acid-base reaction,
       but a **diprotic acid** can transfer two H^+ ions (protons) per molecule, and a **triprotic acid** can
       transfer three H^+ ions per molecule in acid-base reactions. Complete the following reactions
       involving a diprotic acid, and label the acids, bases, conjugate acids and conjugate bases.

   a. $H_2CO_3(aq) + OH^-(aq) \rightleftharpoons$

   b. $HCO_3^-(aq) + OH^-(aq) \rightleftharpoons$

9.5   In aqueous solution, a **strong acid** (Sec. 9.4) is a substance that transfers very nearly 100% of
       its protons to water. A **weak acid** (Sec. 9.4) is a substance that transfers only a few (less than
       5%) of its protons to water. Strong hydroxide bases dissociate completely in water. Weak
       bases react to a small extent with water to produce a few hydroxide ions.

       In the table below, identify each substance as a strong acid, weak acid, strong base or weak
       base:

Substance	Strong acid	Weak acid	Strong base	Weak base
NH_3				
HCl				
KOH				
$Ca(OH)_2$				
H_2SO_4				
H_2CO_3				

9.6   **Salts** (Sec. 9.5) are compounds made up of positive metal or polyatomic ions, and negative
       nonmetal or polyatomic (except hydroxide) ions. Identify each of the following compounds
       as a salt, base, or acid.

Compound	Acid	Base	Salt
HCl			
NaCl			
H_2SO_4			
NaOH			
$CaBr_2$			
$Ba(OH)_2$			

9.7 Salts dissolved in water are completely dissociated into ions in solution. Write a balanced equation for the dissociation of the following soluble ionic compounds in water.

a. KI

b. Na_3PO_4

c. CaI_2

d. Na_2CO_3

9.8 **Neutralization** (Sec. 10.7) is the reaction between an acid and a hydroxide base to form a salt and water. Complete the following neutralization equations by adding the missing products or reactants. Under each reactant molecule, write acid or base, and under each product molecule, write salt or water.

a. HCl + $NaOH$ →

b. → $CaCl_2$ + $2H_2O$

c. → $Mg_3(PO_4)_2$ + $3H_2O$

9.9 In pure water an extremely small number of molecules transfer protons to form the ions H_3O^+ and OH^-.

$$H_2O + H_2O \rightleftharpoons H_3O^+ + OH^-$$

Equal concentrations of H_3O^+ and OH^- are produced by this self-ionization, and are each equal to 1×10^{-7} M. This value can be used to calculate the **ion product constant** (Sec. 9.7).

Ion product constant for water = $[H_3O^+] \times [OH^-] = (1 \times 10^{-7})(1 \times 10^{-7}) = 1 \times 10^{-14}$

The ion product constant relationship is true for water solutions as well as pure water, and so can be used to calculate the concentration of either H_3O^+ or OH^- if the concentration of the other ion is known.

Using the ion product constant for water, determine the following concentrations:

Given concentrations	Substituted equation	Answer
$[H_3O^+] = 2.4 \times 10^{-6}$ M		$[OH^-] =$
$[OH^-] = 3.2 \times 10^{-8}$ M		$[H_3O^+] =$

9.10 Since the hydronium ion concentrations (measures of solution acidity) in aqueous solutions have a very large range of values, a more practical way to represent acidity is by using the **pH scale** (Sec. 9.8), defined as follows:

$$pH = -\log[H_3O^+]$$

Using your calculator, complete the following pH relationships:

[H_3O^+]	pH	acidic, basic, or neutral
1.0×10^{-4}		
1.0×10^{-9}		
2.8×10^{-3}		
7.9×10^{-8}		

9.11 Using the definition of pH and the ion product constant, calculate the following concentrations of ions:

pH	[H_3O^+]	[OH^-]
5.00		
2.00		
3.80		
10.40		

9.12 A **buffer** (Sec. 9.9) solution is a solution that resists a change in pH when small amounts of acid or base are added to it. Buffers (the solutes in buffer solutions) consist of one of the following combinations: a weak acid and the salt of its conjugate base or a weak base and the salt of its conjugate acid. These are known as **conjugate acid-base pairs** (Sec. 9.2).

Predict whether each of the following pairs of substances could function as a buffer in an aqueous solution. Explain your answer.

Pairs of substances	Buffer?	Explanation
KOH, KCl		
H_2CO_3, $NaHCO_3$		
HI, NaI		
NH_4I, NH_3		
KH_2PO_4, H_3PO_4		

9.13   Buffers contain a substance that reacts with and removes added base and a substance that reacts with and removes added acid. Write an equation to show the buffering action in each of the following aqueous solutions.

a. A solution of NH_4I/NH_3 with a small amount of OH^- added.

b. A solution of NaH_2PO_4/H_3PO_4 with a small amount of H_3O^+ added.

c. A solution of NaH_2PO_4/H_3PO_4 with a small amount of OH^- added.

9.14   **Acid-base titration** (Sec. 9.10) is a procedure used to determine the concentration of an acid or base solution. A measured volume of an acid or a base of known concentration is exactly reacted with a measured volume of a base or an acid of unknown concentration. The unknown concentration can be calculated using dimensional analysis.

a. Determine the molarity of an unknown HCl solution, if 21.9 mL of 0.338 M NaOH was needed to neutralize 41.6 mL of the HCl solution. (Hint: Write the balanced neutralization reaction equation.)

b. What is the molarity of an unknown sulfuric acid solution, if 33.2 mL of 0.225 M NaOH is needed to neutralize 13.8 mL of the H_2SO_4 solution? (Hint: Write the balanced neutralization reaction equation.)

Self-Test

True-false: Indicate whether the following statements are true or false. If the statement is false, give the word or phrase that may be substituted for the underlined portion to make the statement true.

1. The value of the ion product constant of water is always <u>1×10^{-10}</u>.
2. Arrhenius defined a base as a substance that, in water, produces <u>hydroxide ions</u>.
3. According the Brønsted-Lowry theory, NH_3 is <u>an acid</u>.
4. Aqueous solutions of acids have a hydronium ion concentration <u>less than</u> 1×10^{-7} moles per liter.
5. A diprotic acid can transfer <u>two protons</u> per molecule during an acid-base reaction.
6. The pH of an acid is the negative logarithm of the <u>hydronium ion concentration</u>.

7. A neutralization reaction produces a salt and <u>a base</u>.

8. <u>Sulfuric acid</u> is the most widely used industrial chemical in the world.

9. A solution whose hydronium ion concentration is 1.0×10^{-4} has a pH of <u>-4</u>.

10. A buffer is a weak acid plus the salt of its conjugate <u>base</u>.

11. An amphoteric substance can function as an acid or a <u>salt</u>.

Multiple choice:

12. In a weak acid, the $[H_3O^+]$ is 2.5×10^{-5}. What is the $[OH-]$?

   a. 7.5×10^{-10} M   b. 2.5×10^{-9} M   c. 4.0×10^{-10} M

   d. 7.5×10^{-9} M   e. 4.0×10^{-9} M

13. The pH of a solution in which $[H_3O^+] = 1.0 \times 10^{-5}$ is:

   a. -5.0 b. -1.5 c. 5.0 d. 1.5 e. none of these

14. The pH of a solution in which $[H_3O^+] = 2.8 \times 10^{-10}$ is:

   a. 9.6 b. -9.6 c. 9.6×10^{-9}

   d. 3.6 e. 3.6×10^{-11}

15. If the pH of a solution is 4.8, the hydronium ion concentration is:

   a. 4.0×10^{-8} M   b. 8.3×10^{-4} M   c. 3.2×10^{-12} M

   d. 1.6×10^{-5} M   e. none of these

16. Which of the following is a diprotic acid?

   a. NH_3 b. H_2SO_4 c. $Ca(OH)_2$ d. H_3PO_4 e. HI

17. Which of the following is a strong base?

   a. NH_3 b. H_2SO_4 c. $Ca(OH)_2$ d. H_3PO_4 e. HI

18. How many milliliters of 0.512 M HCl would be required to neutralize 35.8 mL of 1.50 M KOH?

   a. 11.3 mL   b. 35.8 mL   c. 105 mL   d. 183 mL   e. none of these

19. The conjugate base of the acid HNO_2 is:

   a. H_2O b. H_3O^+ c. NO_2^- d. OH^- e. none of these

20. The salt K_2CO_3 is produced by the reaction of:

   a. a weak acid with a weak base   b. a weak acid with a strong base

   c. a strong acid with a weak base   d. a strong acid with a strong base

   e. none of these

21. If a small amount of hydroxide ion is added to the buffer H_2CO_3/HCO_3^-, the reaction will produce more:

   a. H_3O^+ b. HCO_3^- c. H_2CO_3 d. H_2 e. none of these

Answers to Practice Exercises

9.1

Property	Acid	Base
has a sour taste	X	
turns blue litmus red	X	
has a slippery feel		X
turns red litmus blue		X
has a bitter taste		X

9.2

Base	Conjugate acid
NH_3	NH_4^+
BrO_3^-	$HBrO_3$
HCO_3^-	H_2CO_3

Acid	Conjugate base
HCO_3^-	CO_3^{2-}
$HClO_2$	ClO_2^-
HNO_3	NO_3^-

9.3

a. $\underset{\text{acid}}{HClO_3(aq)} + \underset{\text{base}}{H_2O(l)} \rightarrow \underset{\text{conjugate base}}{ClO_3^-(aq)} + \underset{\text{conjugate acid}}{H_3O^+(aq)}$

b. $\underset{\text{acid}}{HNO_2(aq)} + \underset{\text{base}}{OH^-(aq)} \rightarrow \underset{\text{conjugate base}}{NO_2^-(aq)} + \underset{\text{conjugate acid}}{H_2O(l)}$

9.4

a. $\underset{\text{Acid}}{H_2CO_3(aq)} + \underset{\text{Base}}{OH^-(aq)} \rightleftharpoons \underset{\substack{\text{Conjugate} \\ \text{acid}}}{HCO_3^-(aq)} + \underset{\substack{\text{Conjugate} \\ \text{base}}}{H_2O(l)}$

b. $\underset{\text{Acid}}{HCO_3^-(aq)} + \underset{\text{Base}}{OH^-(aq)} \rightleftharpoons \underset{\substack{\text{Conjugate} \\ \text{acid}}}{CO_3^{2-}(aq)} + \underset{\substack{\text{Conjugate} \\ \text{base}}}{H_2O(l)}$

9.5

Substance	Strong acid	Weak acid	Strong base	Weak base
NH_3				X
HCl	X			
KOH			X	
$Ca(OH)_2$			X	
H_2SO_4	X			
H_2CO_3		X		

9.6

Compound	Acid	Base	Salt
HCl	X		
NaCl			X
H_2SO_4	X		
NaOH		X	
$CaBr_2$			X
$Ba(OH)_2$		X	

9.7   a. $KI \rightarrow K^+ + I^-$

b. $Na_3PO_4 \rightarrow 3Na^+ + PO_4^{3-}$

c. $CaI_2 \rightarrow Ca^{2+} + 2I^-$

d. $Na_2CO_3 \rightarrow 2Na^+ + CO_3^{2-}$

9.8   a. $\underset{\text{acid}}{HCl} + \underset{\text{base}}{NaOH} \rightarrow \underset{\text{salt}}{NaCl} + \underset{\text{water}}{H_2O}$

b. $\underset{\text{acid}}{2HCl} + \underset{\text{base}}{Ca(OH)_2} \rightarrow \underset{\text{salt}}{CaCl_2} + \underset{\text{water}}{2H_2O}$

c. $\underset{\text{acid}}{2H_3PO_4} + \underset{\text{base}}{3Mg(OH)_2} \rightarrow \underset{\text{salt}}{Mg_3(PO_4)_2} + \underset{\text{water}}{3H_2O}$

9.9

Given concentrations	Substituted equation	Answer
$[H_3O^+] = 2.4 \times 10^{-6}$	$[OH^-] = \dfrac{1.00 \times 10^{-14}}{[2.4 \times 10^{-6}]} =$	$[OH^-] = 4.2 \times 10^{-9}$
$[OH^-] = 3.2 \times 10^{-8}$	$[H_3O^+] = \dfrac{1.00 \times 10^{-14}}{[3.2 \times 10^{-8}]} =$	$[H_3O^+] = 3.1 \times 10^{-7}$

9.10

$[H_3O^+]$	pH	acidic, basic, or neutral
1.0×10^{-4}	4.00	acidic
1.0×10^{-9}	9.00	basic
2.8×10^{-3}	2.55	acidic
7.9×10^{-8}	7.10	basic

9.11

pH	$[H_3O^+]$	$[OH^-]$
5.00	1×10^{-5}	1×10^{-9}
2.00	1×10^{-2}	1×10^{-12}
3.80	1.6×10^{-4}	6.3×10^{-11}
10.40	4.0×10^{-11}	2.5×10^{-4}

9.12

Pairs of substances	Buffer?	Explanation
KOH, KCl	No	Strong base and its salt
H_2CO_3, $NaHCO_3$	Yes	Weak acid and the salt of its conjugate base
HI, NaI	No	Strong acid and its salt
NH_4I, NH_3	Yes	Weak base and the salt of its conjugate acid
KH_2PO_4, H_3PO_4	Yes	Weak acid and the salt of its conjugate base

9.13 a. A solution of NH_4I/NH_3 with a small amount of OH^- added.

$$NH_4^+ + OH^- \rightarrow NH_3 + H_2O$$

b. A solution of NaH_2PO_4/H_3PO_4 with a small amount of H_3O^+ added.

$$H_2PO_4^- + H_3O^+ \rightarrow H_3PO_4 + H_2O$$

c. A solution of NaH_2PO_4/H_3PO_4 with a small amount of OH^- added.

$$H_3PO_4 + OH^- \rightarrow H_2PO_4^- + H_2O$$

9.14 a. $HCl + NaOH \rightarrow NaCl + H_2O$

$$M\ HCl = 21.9\ mL\ NaOH \times \frac{0.338\ mole\ NaOH}{1000\ mL\ NaOH} \times \frac{1\ mole\ HCl}{1\ mole\ NaOH} \times \frac{1000\ mL}{0.0416\ L\ HCl} = 0.178\ M\ HCl$$

b. $H_2SO_4 + 2NaOH \rightarrow Na_2SO_4 + 2H_2O$

$$M\ H_2SO_4 = 33.2\ mL\ NaOH \times \frac{0.225\ mole\ NaOH}{1000\ mL\ NaOH} \times \frac{1\ mole\ H_2SO_4}{2\ moles\ NaOH} \times \frac{1000\ mL}{0.0138\ L\ H_2SO_4} = 0.271\ M\ H_2SO_4$$

Answers to Self-Test

The numbers in parentheses refer to sections in your textbook:
1. F; 1.00×10^{-14} (9.7) 2. T (9.1) 3. F; base (9.2) 4. F; greater than (9.8) 5. T (9.3) 6. T (9.8)
7. F; water (9.6) 8. T (9.4) 9. F; 4 (9.8) 10. T (9.9) 11. F; base (9.2) 12. c (9.7) 13. c (9.8)
14. a (9.8) 15. d (9.8) 16. b (9.3) 17. c (9.4) 18. c (9.6) 19. c (9.2) 20. b (9.5) 21. b (9.9)

Chapter Overview

Carbon compounds are the basis of life on Earth; all organic materials are carbon-based. The hydrocarbons, which you will study in this chapter, make up petroleum and are thus an important part of the industrial world, as fuel and in the manufacture of synthetic materials.

In this chapter you will find out how carbon forms such a vast variety of compounds. You will write structural and condensed structural formulas for alkanes and cycloalkanes and name them according to the IUPAC rules. You will identify and draw structural isomers and *cis-trans* isomers. You will write equations for the two major reactions of hydrocarbons.

Practice Exercises

10.1   **Organic chemistry** (Sec. 10.1) is the study of **hydrocarbons** (compounds of hydrogen and carbon, Sec. 10.3) and **hydrocarbon derivatives** (compounds that contain carbon and hydrogen and one or more additional elements, Sec. 10.3).

Hydrocarbons may be **saturated** (containing only single carbon-to-carbon bonds, Sec. 10.4), or **unsaturated** (containing only single carbon-to-carbon multiple bonds, Sec. 10.4). The simplest saturated hydrocarbonds are the **alkanes** (Sec. 10.4). Alkanes may be **cyclic** (carbon atoms arranged in a ring) or **acyclic** (not cyclic).

Methane, ethane, propane, butane, and pentane are acyclic alkanes containing one, two, three, four, and five carbon atoms respectively. Keeping in mind that each carbon atom forms four covalent bonds, complete the following table.

Alkane	Molecular formula	Total number of atoms	Number of C–H bonds	Number of C–C bonds
methane				
ethane				
propane				
butane				
pentane				

10.2   The structures of alkanes and other organic compounds are usually represented in two dimensions, rather than three. Three common types of representations are: **expanded structural formula** (Sec. 10.5), which shows all atoms and all bonds; **condensed structural formula** (Sec. 10.5), which shows groupings of atoms; and skeletal formula, which shows carbon atoms and bonds, but omits hydrogen atoms.

Complete the following table of structural representations. All carbons are connected in a straight chain.

Molecular formula	Condensed structural formula	Expanded structural formula	Skeletal formula

C_2H_6		H H
		  | |
		H−C−C−H
		  | |
		H H	

C_4H_{10}			
	$CH_3-CH_2-CH_2-CH_2-CH_3$		
			C−C−C−C−C−C

10.3 **Structural isomers** (Sec. 10.6) are compounds with the same molecular formula but different structural formulas. Use condensed structural formulas to represent the five isomers of hexane (C_6H_{14}). Hint: One of these is a **continuous-chain alkane** and the others are **branched-chain alkanes** (Sec. 10.6).

10.4 **Conformations** (Sec. 10.7) are differing orientations of a molecule made possible by rotation about a single bond. They are not isomers since one form can change to another without breaking or forming bonds.

Using skeletal formulas, draw two conformations of the straight chain alkane, heptane, C_7H_{16}.

10.5 The IUPAC rules for naming organic compounds make it possible to give each compound a name that uniquely identifies it, and also to draw its structural formula from that name.

For each of the structural formulas below: a) Find the number of carbons in the longest continuous carbon chain. b) Give the IUPAC name. Use the rules for nomenclature found in section 10.8 in your textbook.

CH_3 CH_3 $\mid$ $\mid$ $CH_2-CH_2-CH-CH_3$ a. b.	CH_3 CH_3 $\mid$ $\mid$ $CH_3-CH_2-CH_2-CH-CH-CH_3$ a. b.

CH_3
               $\mid$
CH_3 CH_2 CH_3
$\mid$ $\mid$ $\mid$
$CH_2-CH_2-CH_2-CH-CH-CH_3$
 	
a.
 	
b.
	CH_3
	      $\mid$
	      CH_2
	CH_3 CH_2 CH_3 CH_3
	$\mid$ $\mid$ $\mid$ $\mid$
	$CH_2-CH-CH_2-CH-CH-CH_2-CH_3$
	
	a.
	
	b.

10.6 Once you have learned the IUPAC rules for naming simple alkanes, you can translate the name of an alkane into a structural formula.

Give the condensed structural formulas for the following alkanes:

     a. 3-methylhexane	 b. 2,2-dimethylpentane
     c. 3,5-dimethylheptane	 d. 2,2,4,4-tetramethylhexane

10.7 In a **cycloalkane** (Sec. 10.9), the carbon atoms are attached to one another in a ring-like arrangement. The general formula for cycloalkanes is C_nH_{2n}. IUPAC naming procedures for cycloalkanes are found in Section 10.10 of your textbook. **Line-angle drawings** (Sec. 10.9) are often used to represent cycloalkane structures.

Complete the table below by giving names and molecular formulas and drawing representations with line-angle drawings of the given alkanes.

Name	Molecular formula	Line-angle drawings
methylcyclobutane		
	C_3H_6	▷
1,2-dimethylcyclopentane		
		(cyclobutane with CH_2-CH_3 and CH_3 substituents)
isopropylcyclopropane		

10.8   The problems in this exercise give a review of structural isomerism in alkanes. Compare the following pairs of molecules. Use line-angle notation to draw structural formulas for cycloalkanes and skeletal formulas for acyclic alkanes. Below each structural formula write the molecular formula for the alkane. Tell whether the molecules in each pair are isomers of one another.

a. 1-methylcyclohexane and 1,3-dimethylcyclobutane

b. 2,3-dimethylpentane and 2,2,3-trimethylbutane

c. ethylcyclopropane and 2-methylbutane

10.9 *Cis-trans* **isomers** (Sec. 10.11) are compounds that have the same molecular and structural formulas, but have different spatial arrangements of their atoms because rotation is restricted around bonds.

Give the IUPAC names for the isomers below:

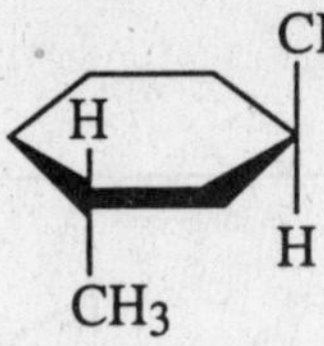

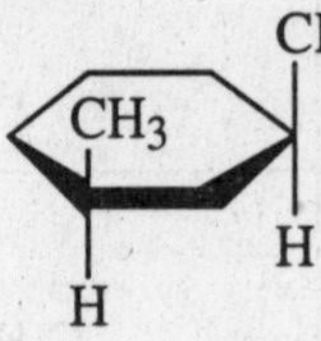

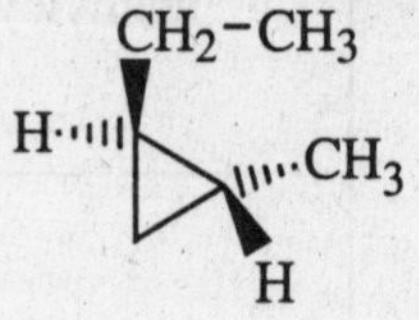

10.10 a. Draw a cyclohexane ring with 2 methyl groups, one on carbon-1 and one on carbon-2.
    b. Draw a cyclohexane ring with 2 methyl groups on carbon-1.

Determine whether *cis-trans* isomerism is possible for structures a. and/or b. Give the IUPAC name for structures a. and b.

a.	b.

10.11 One of the principal reactions of hydrocarbons is **combustion** (Sec. 10.14). Complete combustion is the reaction of a hydrocarbon with oxygen to produce carbon dioxide and water. Write a balanced equation for the complete combustion of the following alkanes:

a   ethane

b. propane

10.12 The other important reaction of alkanes is **halogenation** (Sec. 10.14), the replacement of one or more hydrogen atoms by a halogen (Group VIII element) atom.

a. Draw the product of the monochlorination of cyclopropane.
b. Draw the products (there are three possible) of the monochlorination of hexane.

a.	b.
b.	b.

Self-Test

True-false: Indicate whether the following statements are true or false. If the statement is false, give the word or phrase that may be substituted for the underlined portion to make the statement true.

1. The smallest alkane is <u>ethane</u>.

2. Alkyl groups <u>cannot rotate</u> around the single bonds between the carbons in the ring of a cycloalkane.

3. Straight chain pentane would have a <u>higher boiling point</u> than its structural isomer, 2,2-dimethylpropane.

4. Carbon atoms can form compounds containing chains and rings because carbon has <u>six</u> valence electrons.

5. A hydrocarbon contains only <u>carbon, hydrogen, and oxygen</u> atoms.

6. <u>A saturated</u> hydrocarbon contains at least one carbon-carbon double bond or triple bond.

7. Compounds with the same molecular formulas but different structural formulas are called <u>conformations</u>.

8. <u>An acyclic</u> hydrocarbon contains carbon atoms arranged in a ring structure.

9. The alkyl group $-CH_2-CH_2-CH_3$ is named <u>propanyl</u>.

10. It takes a minimum of <u>three</u> carbon atoms to form a cyclic arrangement of carbon atoms.

11. The formula for a <u>cycloalkane</u> is C_nH_{2n}.

12. Natural gas consists mainly of <u>ethane</u>.

13. Alkanes are good preservatives for metals because they have <u>low boiling points</u>.

14. The reaction of alkanes with oxygen to form carbon dioxide and water is a <u>substitution</u> reaction.

15. The IUPAC name for isopropyl chloride is <u>2-chloropropane</u>.

16. Organic chemistry is defined as the study of <u>hydrocarbons</u>.

Multiple choice:

17. The straight chain alkane having the molecular formula C_6H_{14} is called:

    a. hexane   b. heptane   c. pentane
    d. nonane   e. none of these

18. Using IUPAC rules for naming, if the parent compound is pentane, an ethyl group could be attached to which carbon:

    a. 1   b. 2   c. 3   d. 4   e. 5

19. How many structural isomers can butane have?

    a. 2   b. 3   c. 4   d. 5   e. 6

20. Cyclopropane has the following molecular formula:

    a. CH_4 b. C_2H_6 c. C_3H_8 d. C_4H_{10} e. none of these

21. Which of the compounds below is an isomer of hexane?

   a. methylcyclopentane   b. 2-methylbutane   c. 3-ethylpentane
   d. 2-methylpentane   e. none of these

22. Which of the following compounds forms *cis-trans* isomers?

   a. 2,3-dimethylpentane   b. 2,2-dimethylpentane
   c. 1,1-dimethylcyclopentane   d. 1,2-dimethylcyclopentane
   e. none of these

23. Which of these compounds can have structural isomers?

   a. CH_3Cl b. C_3H_7Cl c. C_3H_8 d. C_2H_5Cl e. none of these

24. Which of these is a correct IUPAC name?

   a. 2-methylcyclobutane   b. *cis*-2,3-dimethylpentane
   c. 1-methylbutane   d. 3-ethylhexane
   e. *cis*-1,2-methypropane

25. How many possible isomers can be written for dimethylcyclobutane?

   a. 2   b. 3   c. 4   d. 5   e. 6

26. Draw structures for the following compounds:

2-bromo-3-methylbutane	*trans*-1,2-dichlorocyclopropane

Answers to Practice Exercises

10.1

Alkane	Molecular formula	Total number of atoms	Number of C–H bonds	Number of C–C bonds
methane	CH_4	5	4	0
ethane	C_2H_6	8	6	1
propane	C_3H_8	11	8	2
butane	C_4H_{10}	14	10	3
pentane	C_5H_{12}	17	12	4

10.2

Molecular formula	Condensed structural formula	Expanded structural formula	Skeletal formula
C_2H_6	CH_3-CH_3	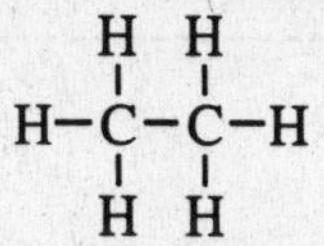 H H | | H-C-C-H | | H H	C—C
C_4H_{10}	$CH_3-CH_2-CH_2-CH_3$	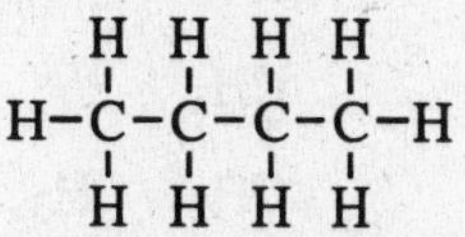 H H H H | | | | H-C-C-C-C-H | | | | H H H H	C—C—C—C
C_5H_{12}	$CH_3-CH_2-CH_2-CH_2-CH_3$	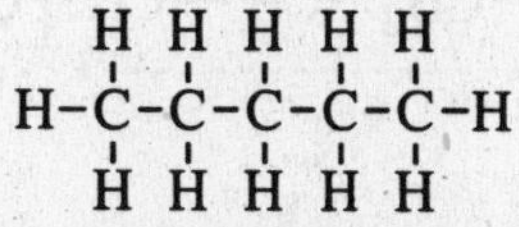 H H H H H | | | | | H-C-C-C-C-C-H | | | | | H H H H H	C—C—C—C—C
C_6H_{14}	$CH_3-CH_2-CH_2-CH_2-CH_2-CH_3$	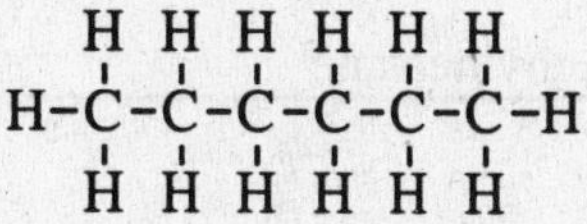 H H H H H H | | | | | | H-C-C-C-C-C-C-H | | | | | | H H H H H H	C—C—C—C—C—C

10.3

$CH_3-CH_2-CH_2-CH_2-CH_2-CH_3$	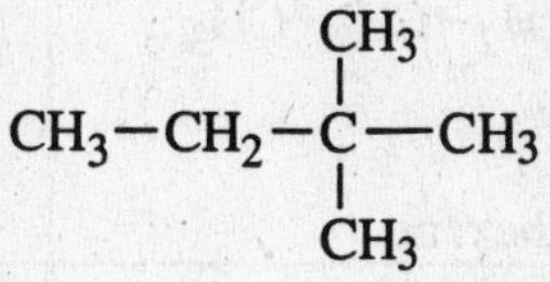 $CH_3-CH_2-\overset{\displaystyle CH_3}{\underset{\displaystyle CH_3}{C}}-CH_3$	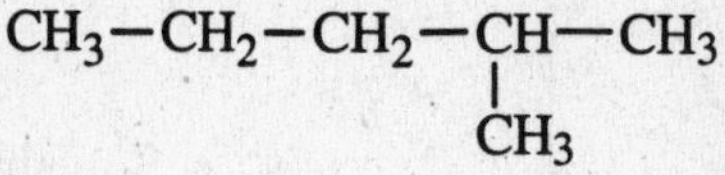 $CH_3-CH_2-CH_2-\underset{\displaystyle CH_3}{CH}-CH_3$
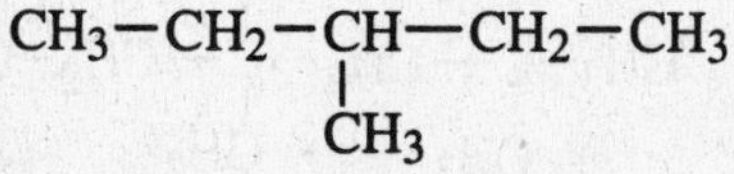 $CH_3-CH_2-\underset{\displaystyle CH_3}{CH}-CH_2-CH_3$	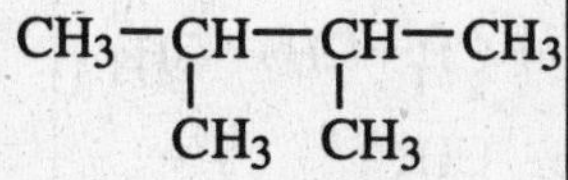 $CH_3-\underset{\displaystyle CH_3}{CH}-\underset{\displaystyle CH_3}{CH}-CH_3$	

10.4 Below are two possible conformations. There are many others.

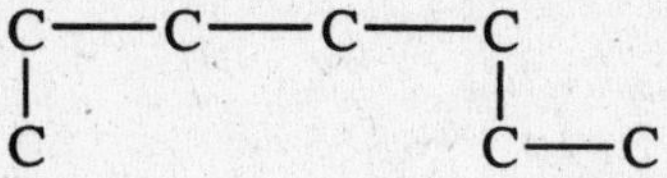

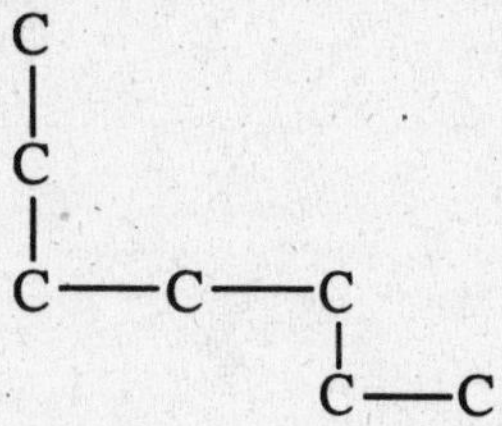

C—C—C—C
| |
C   C—C
	C
	|
	C
	|
	C—C—C
	       |
	       C—C

10.5

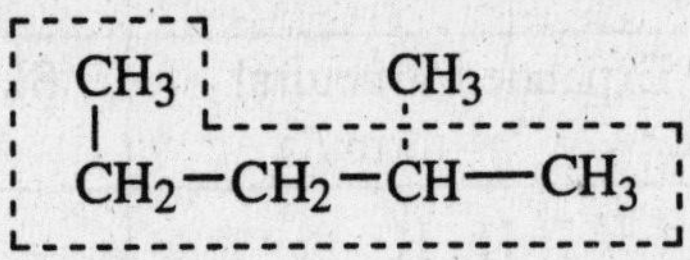

$$CH_2-CH_2-CH-CH_3$$
$$CH_3 \qquad CH_3$$

a. 5 carbon atoms

b. 2-methylpentane

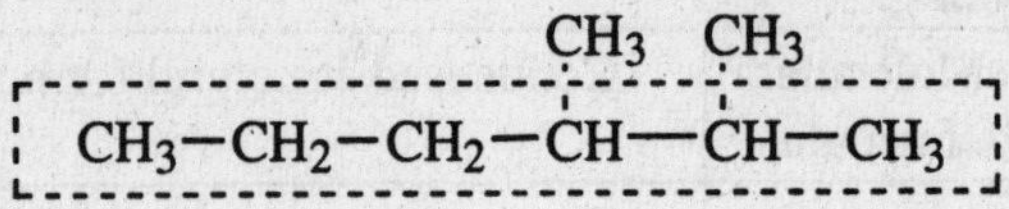

$$CH_3-CH_2-CH_2-CH-CH-CH_3$$
$$CH_3 \quad CH_3$$

a. 6 carbon atoms

b. 2,3-dimethylhexane

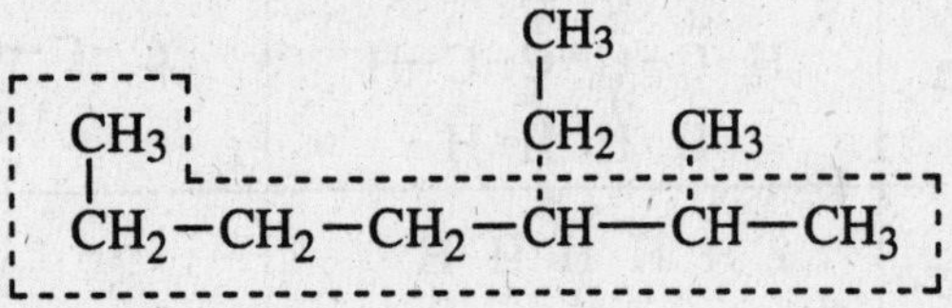

$$CH_2-CH_2-CH_2-CH-CH-CH_3$$
$$CH_3 \qquad CH_2 \quad CH_3$$
$$CH_3$$

a. 7 carbon atoms

b. 3-ethyl-2-methylheptane

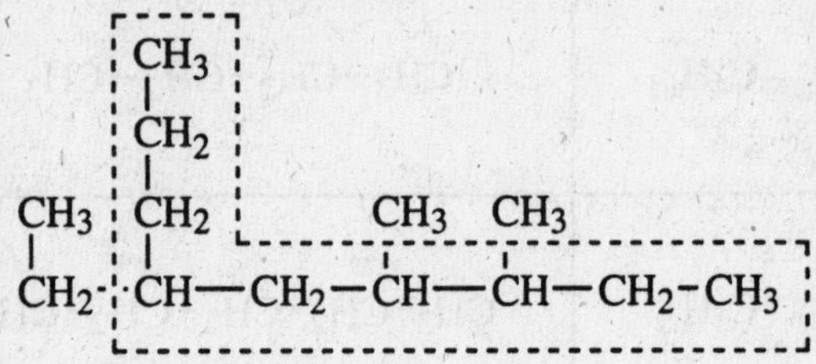

$$CH_2-CH-CH_2-CH-CH-CH_2-CH_3$$
$$CH_3 \quad CH_2 \qquad CH_3 \quad CH_3$$
$$CH_2$$
$$CH_3$$

a. 9 carbon atoms

b. 6-ethyl-3,4-dimethylnonane

10.6

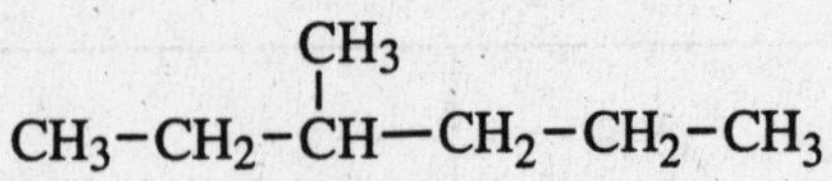

$$CH_3-CH_2-CH-CH_2-CH_2-CH_3$$
$$CH_3$$

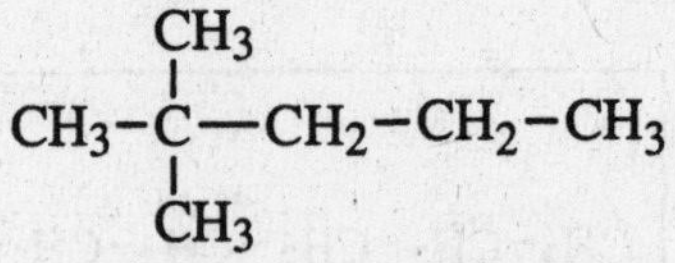

a. 3-methylhexane

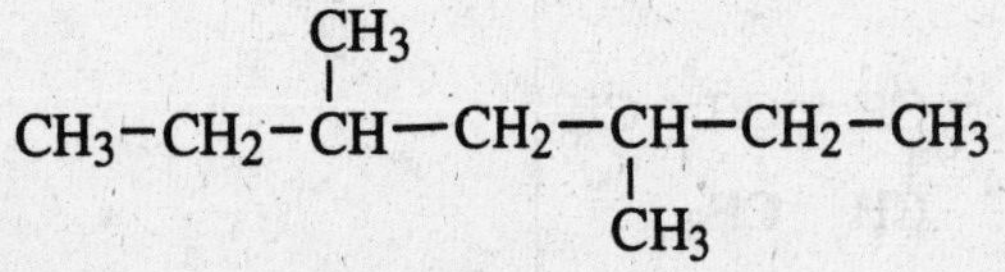

$$CH_3-\underset{\underset{CH_3}{|}}{\overset{\overset{CH_3}{|}}{C}}-CH_2-CH_2-CH_3$$

b. 2,2-dimethylpentane

$$CH_3-CH_2-CH-CH_2-CH-CH_2-CH_3$$
$$CH_3 \qquad\qquad CH_3$$

c. 3,5-dimethylheptane

$$CH_3-\underset{\underset{CH_3}{|}}{\overset{\overset{CH_3}{|}}{C}}-CH_2-\underset{\underset{CH_3}{|}}{\overset{\overset{CH_3}{|}}{C}}-CH_2-CH_3$$

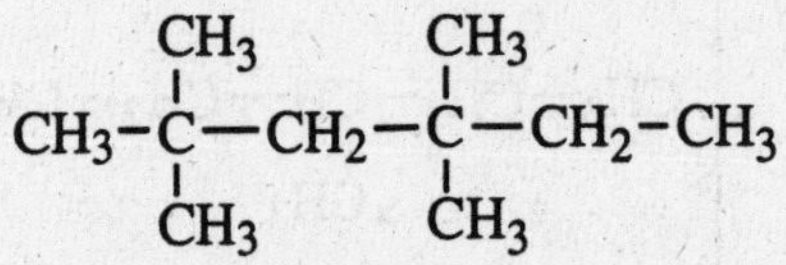

d. 2,2,4,4-tetramethylhexane

10.7

Name	Molecular formula	Line-angle drawings
methylcyclobutane	C_5H_{10}	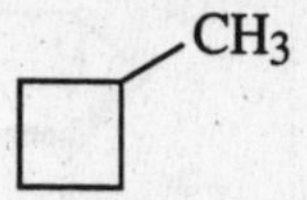
cyclopropane	C_3H_6	
1,2-dimethylcyclopentane	C_7H_{14}	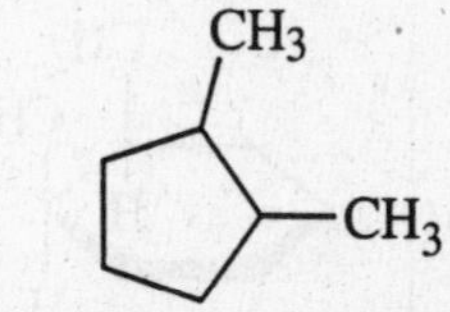
1-ethyl-2-methylcyclobutane	C_6H_{12}	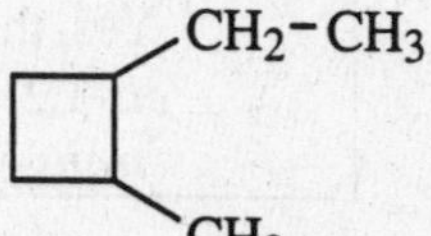
isopropylcyclopropane	C_6H_{12}	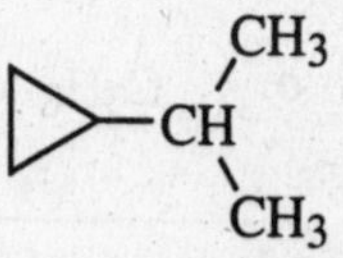

10.8 a.

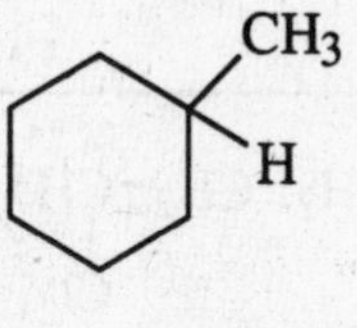

C_7H_{14} 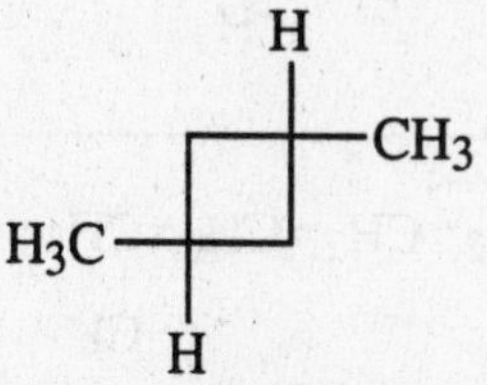 C_6H_{12}

No, these are not isomers of one another because they have different molecular formulas.

b.

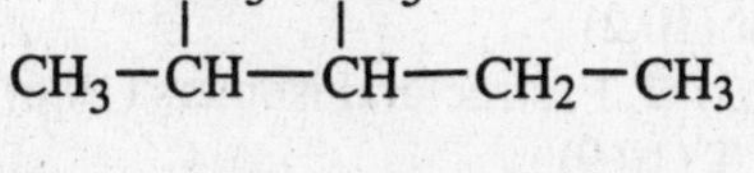

$$CH_3-\overset{\overset{\displaystyle CH_3}{|}}{CH}-\overset{\overset{\displaystyle CH_3}{|}}{CH}-CH_2-CH_3 \qquad CH_3-\overset{\overset{\displaystyle CH_3}{|}}{\underset{\underset{\displaystyle CH_3}{|}}{C}}-\overset{\overset{\displaystyle }{}}{\underset{\underset{\displaystyle CH_3}{|}}{CH}}-CH_3$$

C_7H_{16} 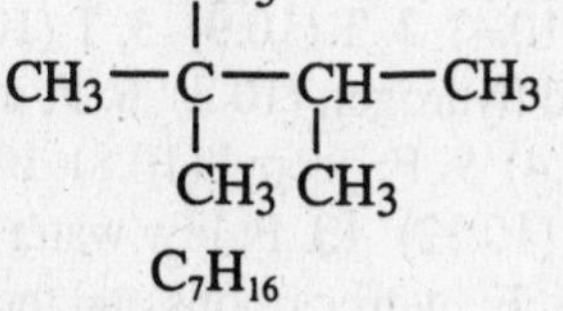 C_7H_{16}

Yes, these are isomers. They have the same molecular formula.

c.

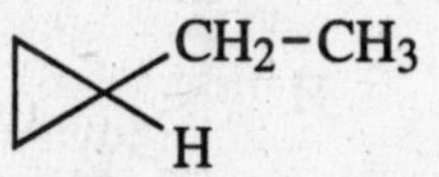

     $CH_3-\overset{\overset{\displaystyle CH_3}{|}}{CH}-CH_2-CH_3$

C_5H_{10} 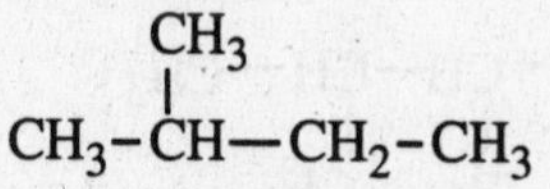 C_5H_{12}

No, these are not isomers of one another. Their molecular formulas differ.

10.9

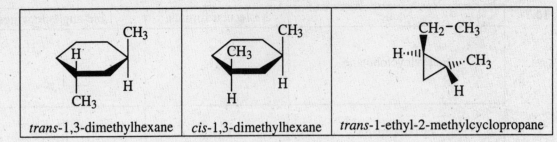

trans-1,3-dimethylhexane   *cis*-1,3-dimethylhexane   *trans*-1-ethyl-2-methylcyclopropane

10.10

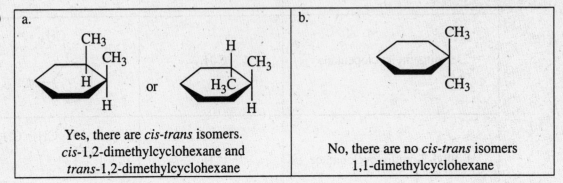

a.

or

Yes, there are *cis-trans* isomers.
cis-1,2-dimethylcyclohexane and
trans-1,2-dimethylcyclohexane

b.

No, there are no *cis-trans* isomers
1,1-dimethylcyclohexane

10.11   a. $2\,C_2H_6 + 7\,O_2 \longrightarrow 4\,CO_2 + 6\,H_2O$

b. $C_3H_8 + 5\,O_2 \longrightarrow 3\,CO_2 + 4\,H_2O$

10.12

a.

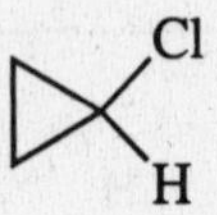

b. $CH_3-CH_2-CH_2-CH_2-CH_2-CH_2$ | Cl

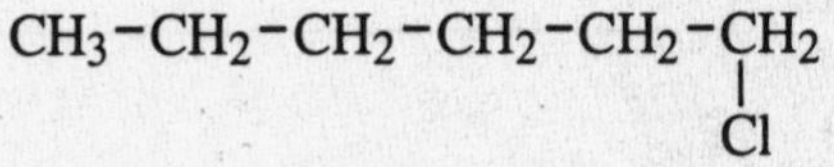

b. $CH_3-CH_2-CH_2-CH_2-CH-CH_3$ | Cl

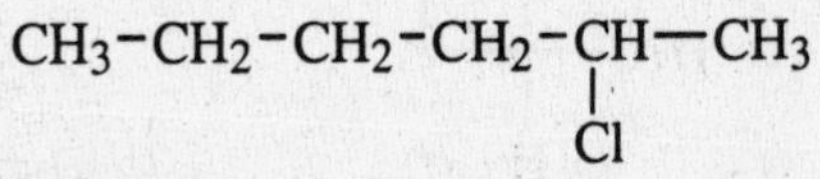

b. $CH_3-CH_2-CH_2-CH-CH_2-CH_3$ | Cl

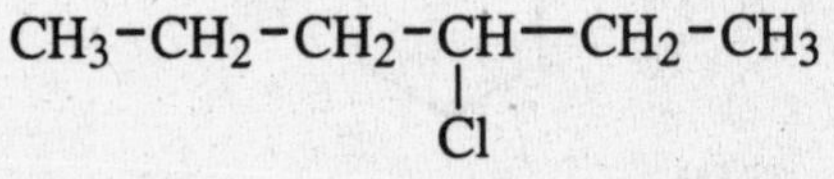

Answers to Self-Test

The numbers in parentheses refer to sections in your textbook:
1. F; methane (10.4) **2.** T (10.9) **3.** T (10.13) **4.** F; four (10.2)
5. F; carbon and hydrogen (10.3) **6.** F; unsaturated (10.3) **7.** F; structural isomers (10.6)
8. F; cyclic (10.4) **9.** F; propyl (10.8) **10.** T (10.9) **11.** T (10.9)
12. F; methane (10.12) **13.** F; low water solubility (10.13) **14.** F; combustion (10.14)
15. T (10.8) **16.** F; hydrocarbons and their derivatives (10.1) **17.** a (10.8) **18.** c (10.8) **19.** a (10.6)
20. e; C_3H_6 (10.9) **21.** d (10.6) **22.** d (10.11) **23.** b (10.6) **24.** d (10.8, 10.11) **25.** d (10.9, 10.11)

26.

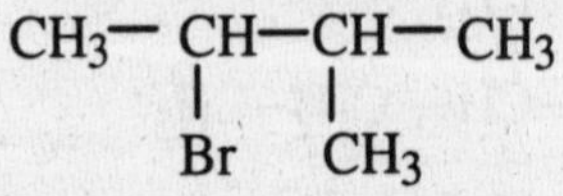

$CH_3-CH-CH-CH_3$ | Br | CH_3

2-bromo-3-methylbutane

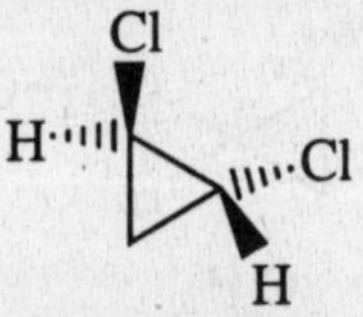

trans-1,2-dichlorocyclopropane

Chapter Overview

Unsaturated hydrocarbons contain fewer than the largest possible number of hydrogen atoms since they have one or more carbon-carbon double or triple bonds. Because a multiple bond is more easily broken than a single bond, unsaturated hydrocarbons are more reactive chemically than saturated hydrocarbons.

In this chapter you will name and write structural formulas for alkenes, alkynes, and aromatic hydrocarbons. You will learn the characteristics of these compounds, their physical and chemical properties, and their most common reactions.

Practice Exercises

11.1 An **alkene** (Sec. 11.2) contains one or more carbon-carbon double bonds. The IUPAC names for alkenes are similar to those for alkanes, with the *-ene* ending replacing the *-ane* ending. The longest carbon chain must contain the double bond. The location of the double bond is indicated with a single number, that of the first carbon atom of the double bond.

Write the IUPAC names for the following alkenes and **cycloalkenes** (Sec. 11.2).

$CH_3-C=CH_2$
         $\vert$
         CH_3
 a.
	$CH_3-CH_2-CH_2-CH-CH=CH_2$
	                           $\vert$
	                           CH_3
	 b.

(cyclohexene with CH_3) c.	(cyclopentadiene with CH_2-CH_3) d.

11.2 Draw structural formulas for the following alkenes and cycloalkenes.

a. 3-methyl-1-pentene	b. 2-methyl-1,3-pentadiene
c. 1-methylcyclopentene	d. 1,4-cycloheptadiene

11.3   *Cis-trans* isomerism is possible for some alkenes, since the double bond is rigid, preventing rotation around its axis. For *cis-trans* isomers to exist, each of the two carbon atoms of the double bond must have two different groups attached to it.

Determine whether each of these alkenes can exist as *cis-trans* isomers. If isomers do exist, draw and give the IUPAC name for each isomer.

$CH_3-C=CH_2$ $\quad\quad\;\; \mid$ $\quad\quad\; CH_3$ a.	$CH_3-CH=CH-CH_2-CH_3$ b.
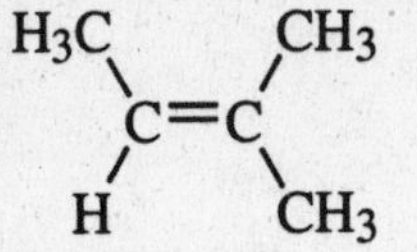 $H_3C\quad\quad CH_3$ $\quad\;\diagdown\quad\diagup$ $\quad\quad C=C$ $\quad\diagup\quad\diagdown$ $H\quad\quad\; CH_3$ c.	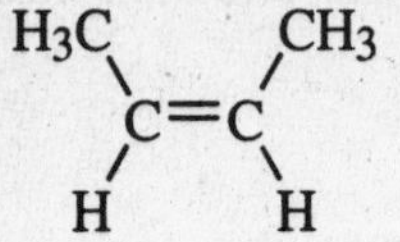 $H_3C\quad\quad CH_3$ $\quad\;\diagdown\quad\diagup$ $\quad\quad C=C$ $\quad\diagup\quad\diagdown$ $H\quad\quad\; H$ d.

11.4   Draw structural formulas for the following alkenes.

a. *trans*-3-hexene	b. *cis*-2-hexene	c. *trans*-3,4-dimethyl-3-heptene

11.5   The most important reactions of alkenes are **addition reactions** (Sec. 11.5). **Symmetrical reaction**s (Sec. 11.5), such as hydrogenation and halogenation, involve adding identical atoms to each carbon of the double bond. In hydrogenation, a hydrogen atom is added to each carbon of the double bond by heating the alkene and H_2 in the presence of a catalyst. Halogenation involves the use of Br_2 or Cl_2 to add a halogen atom to each carbon of the double bond.

Complete the following reactions. Give the structural formula for the product.

a. $CH_3-CH_2-CH=CH_2$ + H_2 $\xrightarrow[\text{catalyst}]{\text{Ni}}$

b. $CH_3-CH_2-CH=CH_2$ + Br_2 $\longrightarrow$

c. 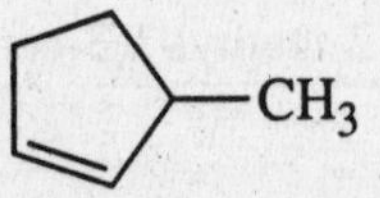   [cyclopentene ring]$-CH_3$ + H_2 $\xrightarrow[\text{catalyst}]{\text{Ni}}$

11.6   Addition reactions may also be **unsymmetrical** (Sec. 11.5); different atoms or groups of atoms are added to the carbons of the double bond. Hydrohalogenation and hydration are important types of unsymmetrical addition. **Markovnikov's rule** (Sec. 11.5) states that, in an unsymmetrical addition, the hydrogen atom from the molecule being added becomes attached to the unsaturated carbon atom that already has the most hydrogen atoms.

Complete the following reactions. In each case give the structural formula for the major expected product.

a. $CH_3-CH_2-CH=CH_2$ + HBr $\longrightarrow$

b. $CH_3-CH_2-CH=CH_2$ + H_2O $\xrightarrow{H_2SO_4}$

c. 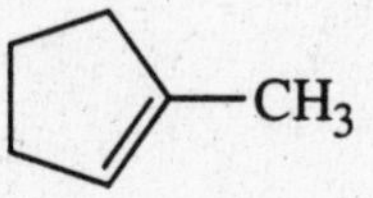 (cyclopentene)—CH_3 + HBr $\longrightarrow$

d. 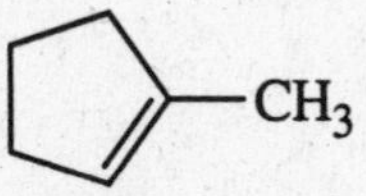 (cyclopentene)—CH_3 + H_2O $\xrightarrow{H_2SO_4}$

11.7 Write the name of the alkene that could be used to prepare each of the following compounds. Remember Markovnikov's rule.

$CH_3-CH_2-CH_2-CH-CH_3$ $\quad\quad\quad\quad\quad\quad\;\;\; \vert$ $\quad\quad\quad\quad\quad\quad\; Br$	$CH_3-CH-CH-CH_3$ $\quad\quad\quad\;\;\; \vert \quad\;\; \vert$ $\quad\quad\quad CH_3 \; OH$	(cyclopentane with Br and CH_3)
a.	b.	c.

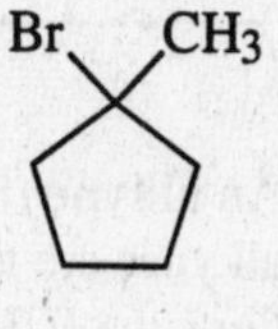

11.8 How many molecules of hydrogen gas, H_2, would react with one molecule of each of the following compounds? Give the structure and IUPAC name for each product.

$CH_2=CH-CH_2-CH=CH_2$	(cyclooctadiene ring)
a.	b.

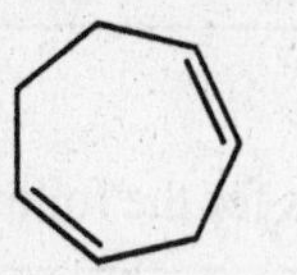

11.9 A **polymer** (Sec. 11.6) is a very large molecule composed of many identical repeating units. Alkenes can form **addition polymers** (Sec. 11.6) when the alkene monomers simply add together. The double bond of the alkene is broken and single carbon-carbon bonds form between monomers.

Draw structural formulas for the monomer units from which these addition polymers were made and for the first three repeating units of the polymers.

General formula of polymer	Monomer formula	First three units of polymer structural formula
$\left(\begin{array}{cc} H & H \\ \mid & \mid \\ -C & -C- \\ \mid & \mid \\ H & F \end{array}\right)_n$		
$\left(\begin{array}{cc} H & H \\ \mid & \mid \\ -C & -C- \\ \mid & \mid \\ H & CH_2 \\ & \mid \\ & CH_3 \end{array}\right)_n$		

11.10 An **alkyne** (Sec. 11.7) has one or more carbon-carbon triple bonds. The rules for naming alkynes are the same as those for naming alkenes, with the ending *-yne* instead of *-ene*. Give the IUPAC names for the following alkynes:

$CH_3-CH_2-CH_2-C\equiv CH$	$CH_3-CH_2-C\equiv C-CH_3$	$CH_3-CH_2-\overset{\overset{\displaystyle CH_3}{\mid}}{\underset{\underset{\displaystyle CH_3}{\mid}}{C}}-C\equiv C-CH_3$
a.	b.	c.

11.11 Draw structural formulas for the following alkynes:

a. 3-methyl-1-pentyne	b. 4-methyl-2-pentyne	c. 4,5-dimethyl-2-hexyne

11.12 Addition reactions of alkynes are similar to those of alkenes. However, two molecules of a specific reactant can add to the triple bond.

Complete the following reactions. Give the structural formula for the major expected product.

a. $CH_3-CH_2-CH_2-C\equiv CH$ + 2 Cl_2 ⟶

b. $CH_3-CH_2-CH_2-C\equiv CH$ + 2 HBr ⟶

c. $CH_3-CH_2-CH_2-C\equiv CH$ + 1 HCl ⟶

d. $CH_3-CH_2-CH_2-C\equiv CH$ + 2 H_2 $\xrightarrow[\text{catalyst}]{\text{Ni}}$

11.13 **Aromatic hydrocarbons** (Sec. 11.8) are unsaturated cyclic hydrocarbons whose bonding is different from ordinary double or triple bonds. All of the carbon-to-carbon bonds in the aromatic structure are equal and less easily broken than ordinary multiple bonds. The naming of the compounds below is based on the aromatic hydrocarbon benzene. The name of the substituent on the benzene ring is used as a prefix.

Give the IUPAC name for the following aromatic compounds:

a.

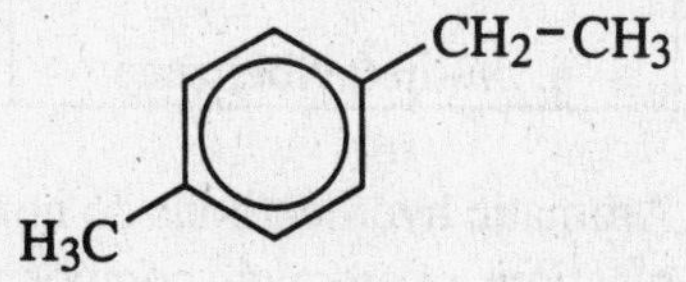

b.

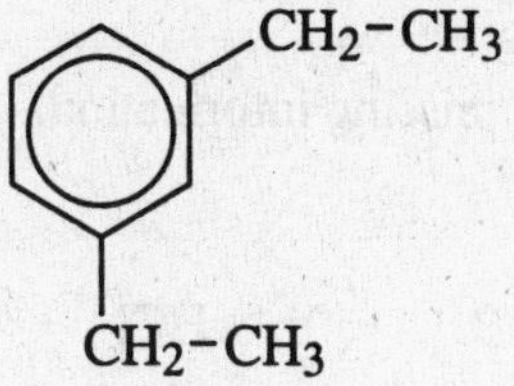

c.

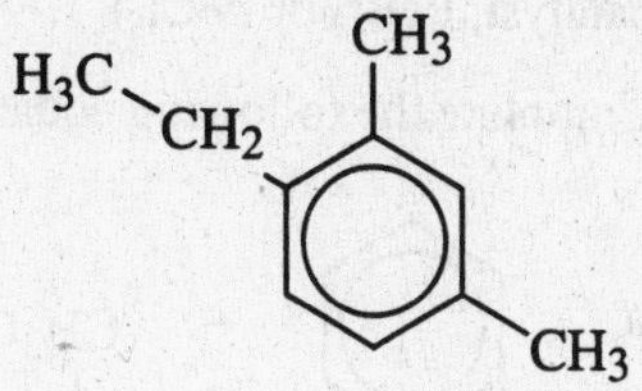

d.

11.14 When there are two substituents on the benzene ring, the molecule can be named using the prefix system (Sec. 11.13). Name the molecules in the b. and c. parts of the exercise above using the prefix system.

b.

c.

11.15 If the group attached to the benzene ring is not easily named as a substituent, the benzene ring is treated as the attached group and is called a *phenyl* group. The compound is then named as an alkane, an alkene, or an alkyne.

Write the IUPAC name for the following phenyl-substituted compounds:

$CH_3-CH_2-CH_2-CH-CH_3$ a.	$CH_3-CH=CH-CH-CH_3$ b.

11.16 Draw structural formulas for the following compounds:

a. *m*-diethylbenzene	b. 1-ethyl-4-methylbenzene	c. 3-phenyl-1-hexene

11.17 Aromatic hydrocarbons do not readily undergo addition reactions. The most important reactions of aromatic compounds are substitution reactions. These include alkylation (using alkyl halides and the catalyst $AlCl_3$), and halogenation (using Br_2 or Cl_2 in the presence of a catalyst, $FeBr_3$ or $FeCl_3$).

Complete the following equations by supplying the missing information.

a. ⬡ + CH_3Cl $\xrightarrow{AlCl_3}$? + HCl

b. ? + ? $\xrightarrow{FeBr_3}$ ⬡Br + HBr

c. ⬡ + Cl_2 $\xrightarrow{FeCl_3}$? + ?

d. ? + ? $\xrightarrow{\text{?}}$ ⬡CH_2-CH_3 + HCl

11.18 Use this identification exercise to review your knowledge of the structures introduced in this chapter. Using structures A through I, give the best choice for each of the terms below.

CH_3 H C=C H CH_3 **A.**	H H C=C H_3C CH_3 **B.**	$CH_3-CH_2-CH_2-CH_3$ **C.**
H_3C-⬡ **D.**	CH_3 ⬡CH_3 **E.**	$CH_3-CH_2-C{\equiv}CH$ **F.**

a. aromatic compound _____
b. isomer of compound B _____
c. alkyne _____
d. cyclic alkane _____
e. *cis*-isomer of an alkene _____
f. straight chain alkane _____

Give the IUPAC names for structures A through F above:

g. structure A _____
h. structure B _____
i. structure C _____
j. structure D _____
k. structure E _____
l. structure F _____

Self-Test

True-false: Indicate whether the following statements are true or false. If the statement is false, give the word or phrase that may be substituted for the underlined portion to make the statement true.

1. The specific part of a molecule that governs the molecule's chemical properties is called a <u>functional</u> group.
2. The general formula for an <u>alkene</u> is C_nH_{2n}.
3. When naming alkenes by the IUPAC system, select as the parent carbon chain the longest chain that contains <u>at least one carbon atom</u> of the double bond.
4. The name of a cycloalkene <u>does not include</u> the numbered location of the double bond.
5. The carbon atoms at each end of a double bond have a <u>trigonal planar</u> arrangement of bonds.
6. The units that make up a polymer molecule are called <u>copolymers</u>.

7. Benzene undergoes an <u>addition</u> reaction with bromine in the presence of a catalyst.

8. <u>Propene</u> is the simplest alkene that has *cis-trans* isomerism.

9. If 2-butene is halogenated with bromine gas, the most probable product is <u>2-bromobutane</u>.

10. Hydration of an alkene produces <u>an alcohol</u>.

11. Polyethylene contains <u>many</u> double bonds.

Multiple choice:

12. According to Markovnikov's rule, addition of HBr to the double bond of 1-hexene would produce:

    a. 1-bromohexane   b. 1,2-dibromohexane   c. 2-bromo-1-hexene
    d. 2-bromohexane   e. none of these

13. The geometry of the carbon-carbon triple bond of an alkyne is:

    a. linear   b. trigonal planar   c. tetrahedral
    d. angular   e. none of these

14. Which of these is a correct name according to IUPAC rules?

    a. 2-methylbenzene   b. 1-methylbenzene   c. 1-ethyl-2-methylbenzene
    d. 2,4-dimethylbenzene   e. none of these

15. Another name for *meta*-dimethylbenzene is:

    a. 1,2-dimethylbenzene   b. 2,3-dimethylbenzene   c. 1,3-dimethylbenzene
    d. 1,4-dimethylbenzene   e. none of these

16. Aromatic compounds have structures based on what parent molecule?

    a. benzene   b. cyclopropane   c. cyclohexane
    d. hexane   e. none of these

17. Which of the following compounds does **not** have the formula C_5H_8?

    a. 1-methylcyclobutene   b. 3-methyl-1-butyne   c. 2-methyl-1-butene
    d. 1,3-pentadiene   e. none of these

18. If you wished to prepare 2-bromo-3-methylbutane by the addition of HBr to an alkene, which of these alkenes would you use?

    a. 2-methyl-2-butene   b. 2-methyl-1-butene   c. 3-methyl-1-butene
    d. 3-methyl-2-butene   e. none of these

19. Which of the following compounds would be the most likely to undergo an addition reaction with HCl?

    a. toluene   b. cyclohexene   c. ethylbenzene
    d. heptane   e. none of these

Answers to Practice Exercises

11.1   a. 2-methylpropene;
           b. 3-methyl-1-hexene;
           c. 3-methylcyclohexene;
           d. 1-ethyl-1,3-cyclopentadiene

11.2

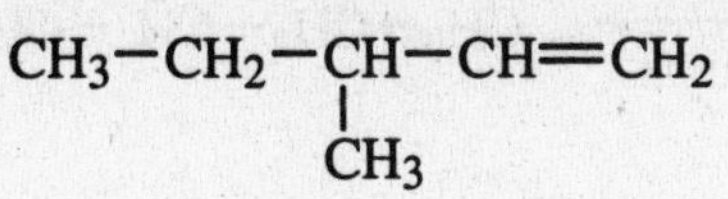

CH$_3$—CH$_2$—CH—CH=CH$_2$ 　　　　　CH$_3$ a. 3-methyl-1-pentene	CH$_2$=C—CH=CH—CH$_3$ 　　　CH$_3$ b. 2-methyl-1,3-pentadiene
—CH$_3$ c. 1-methylcyclopentene	 d. 1,4-cycloheptadiene

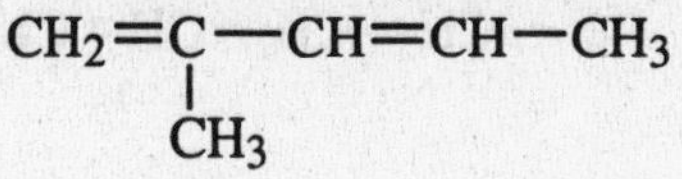

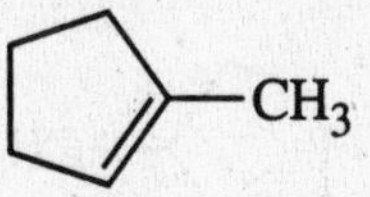

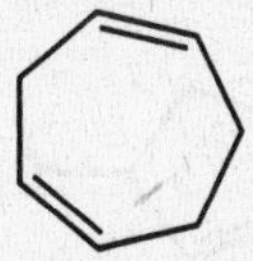

11.3　　a. no

b. yes; *trans*-2-pentene and *cis*-2-pentene

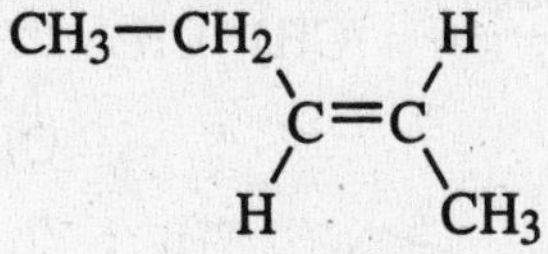

CH$_3$—CH$_2$　　H 　　　C=C 　H　　CH$_3$ *trans*-2-pentene	H　　H 　C=C H$_3$C　　CH$_2$—CH$_3$ *cis*-2-pentene

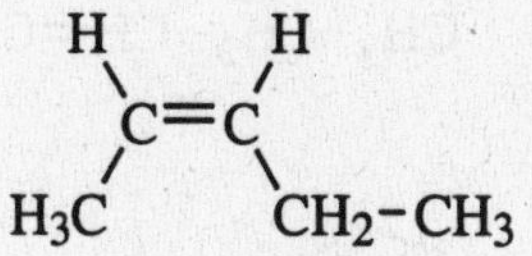

c. no

d. yes; *cis*-2-butene and *trans*-2-butene

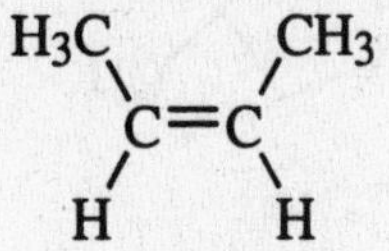

H$_3$C　　CH$_3$ 　C=C H　　H *cis*-2-butene	H$_3$C　　H 　C=C H　　CH$_3$ *trans*-2-butene

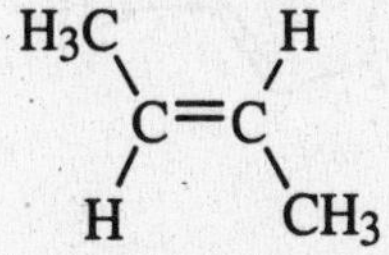

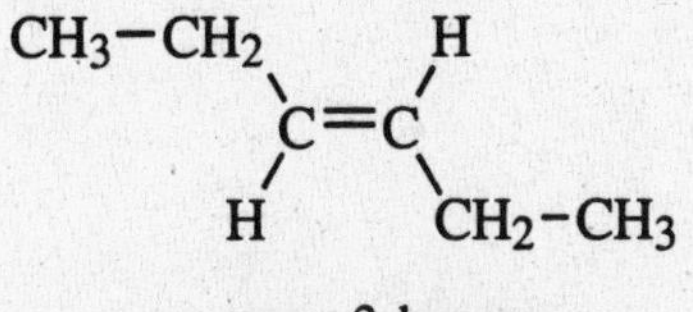

11.4

CH$_3$—CH$_2$　　H 　　　C=C 　H　　CH$_2$—CH$_3$ a. *trans*-3-hexene	H　　H 　C=C H$_3$C　　CH$_2$—CH$_2$—CH$_3$ b. *cis*-2-hexene
CH$_3$—CH$_2$　　CH$_3$ 　　　C=C 　H$_3$C　　CH$_2$—CH$_2$—CH$_3$ c. *trans*-3,4-dimethyl-3-heptene	

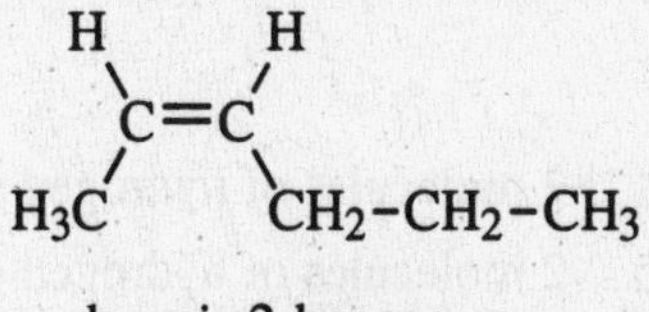

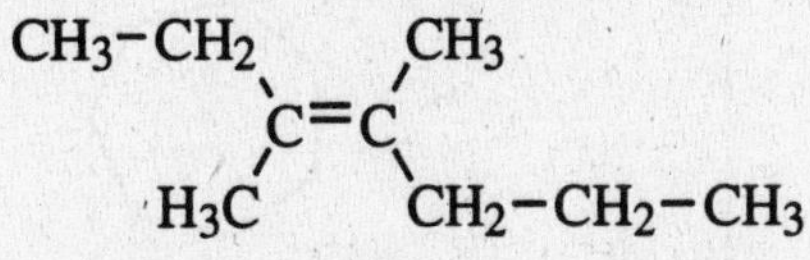

11.5 a. $CH_3-CH_2-CH=CH_2$ + H_2 $\xrightarrow[\text{catalyst}]{\text{Ni}}$ $CH_3-CH_2-CH_2-CH_3$

b. $CH_3-CH_2-CH=CH_2$ + Br_2 $\longrightarrow$ $CH_3-CH_2-\underset{\underset{Br}{|}}{CH}-\underset{\underset{Br}{|}}{CH_2}$

c. 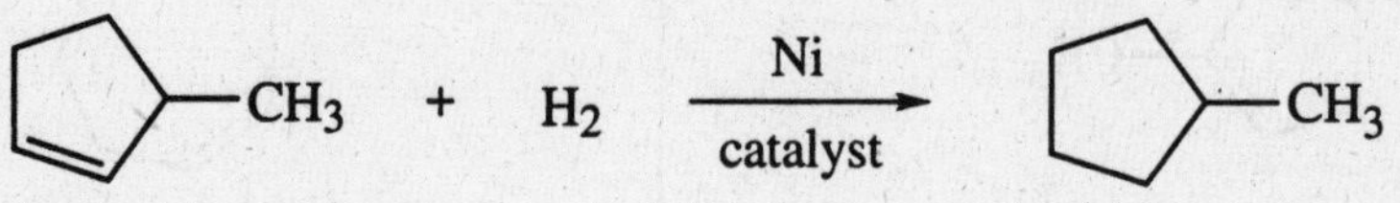 $-CH_3$ + H_2 $\xrightarrow[\text{catalyst}]{\text{Ni}}$ $-CH_3$

11.6 a. $CH_3-CH_2-CH=CH_2$ + HBr $\longrightarrow$ $CH_3-CH_2-\underset{\underset{Br}{|}}{CH}-CH_3$

b. $CH_3-CH_2-CH=CH_2$ + H_2O $\xrightarrow{H_2SO_4}$ $CH_3-CH_2-\underset{\underset{OH}{|}}{CH}-CH_3$

c. 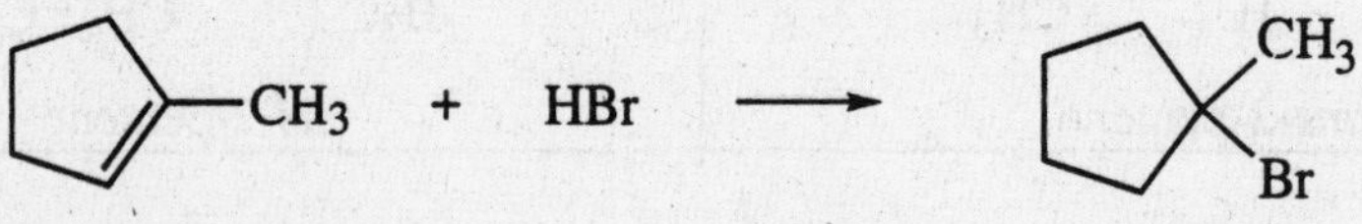 $-CH_3$ + HBr $\longrightarrow$

d. 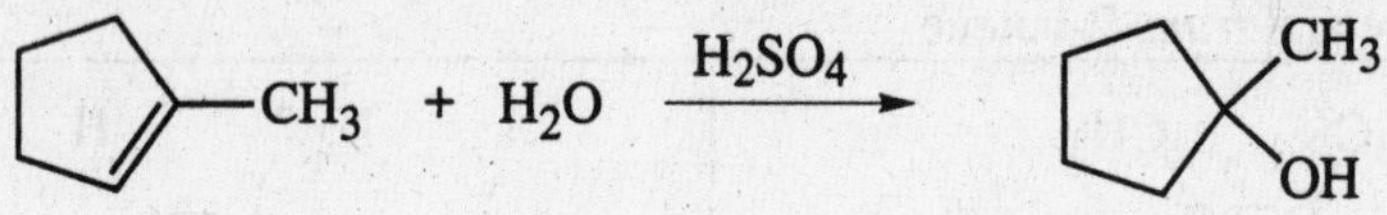 $-CH_3$ + H_2O $\xrightarrow{H_2SO_4}$

11.7 a.   1-pentene
       b.   3-methyl-1-butene
       c.   1-methylcyclopentene

11.8 a.   2 molecules of hydrogen gas.
       b.   2 molecules of hydrogen gas.

$CH_3-CH_2-CH_2-CH_2-CH_3$	
a. pentane	b. cycloheptane

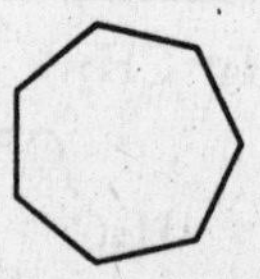

11.9

General formula of polymer	Monomer formula	First three units of polymer structural formula
$\left(\begin{array}{cc} H & H \\ -C-C- \\ H & F \end{array}\right)_n$	$CH_2=CH-F$	$\left(\begin{array}{cccccc} H & H & H & H & H & H \\ -C-C-C-C-C-C- \\ H & F & H & F & H & F \end{array}\right)$
$\left(\begin{array}{cc} H & H \\ -C-C- \\ H & CH_2 \\ & CH_3 \end{array}\right)_n$	$CH_2=CH-CH_2-CH_3$	$\left(\begin{array}{cccccc} H & H & H & H & H & H \\ -C-C-C-C-C-C- \\ H & CH_2 & H & CH_2 & H & CH_2 \\ & CH_3 & & CH_3 & & CH_3 \end{array}\right)$

11.10 a. 1-pentyne
  b. 2-pentyne
  c. 4,4-dimethyl-2-hexyne

11.11

$CH_3-CH_2-\underset{\underset{CH_3}{\mid}}{CH}-C\equiv CH$	$CH_3-\underset{\underset{CH_3}{\mid}}{CH}-C\equiv C-CH_3$	$CH_3-\underset{\underset{CH_3}{\mid}}{CH}-\underset{\underset{CH_3}{\mid}}{CH}-C\equiv C-CH_3$
a. 3-methyl-1-pentyne	b. 4-methyl-2-pentyne	c. 4,5-dimethyl-2-hexyne

11.12 a. $CH_3-CH_2-CH_2-C\equiv CH \quad + \quad 2\,Cl_2 \longrightarrow CH_3-CH_2-CH_2-\underset{\underset{Cl}{\mid}}{\overset{\overset{Cl}{\mid}}{C}}-\underset{\underset{Cl}{\mid}}{\overset{\overset{Cl}{\mid}}{CH}}$

  b. $CH_3-CH_2-CH_2-C\equiv CH \quad + \quad 2\,HBr \longrightarrow CH_3-CH_2-CH_2-\underset{\underset{Br}{\mid}}{\overset{\overset{Br}{\mid}}{C}}-CH_3$

  c. $CH_3-CH_2-CH_2-C\equiv CH \quad + \quad 1\,HCl \longrightarrow CH_3-CH_2-CH_2-\underset{\underset{Cl}{\mid}}{C}=CH_2$

  d. $CH_3-CH_2-CH_2-C\equiv CH \quad + \quad 2\,H_2 \xrightarrow[\text{catalyst}]{Ni} CH_3-CH_2-CH_2-CH_2-CH_3$

11.13 a   benzene
  b. 1-ethyl-4-methylbenzene
  c. 1,3-diethylbenzene
  d. 1-ethyl-2,4-dimethylbenzene

11.14 b. *p*-ethylmethylbenzene
  c. *m*-diethylbenzene

11.15 a. 2-phenylpentane
b. 4-phenyl-2-pentene

11.16

| a. *m*-diethylbenzene | b. 1-ethyl-4-methylbenzene | c. 3-phenyl-1-hexene |

11.17 a.

b.

c.

d.

11.18   a. E; b. A; c. F; d. D; e. B; f. C;
g. *trans*-2-butene;
h. *cis*-2-butene;
i. butane;
j. methylcyclohexane;
k. 1,2-dimethylbenzene or *o*-dimethylbenzene
l. 1-butyne.

Answers to Self-Test

The numbers in parentheses refer to sections in your textbook:
1. T (11.1) **2.** T (11.2) **3.** F; both carbon atoms (11.3) **4.** T (11.3) **5.** T (11.2)
6. F; monomers (11.6) **7.** F; substitution (11.10) **8.** F; 2-butene (11.4)
9. F; 2,3-dibromobutane (11.5) **10.** T (11.5) **11.** F; no (11.6) **12.** d (11.5) **13.** a (11.7)
14. c (11.9) **15.** c (11.9) **16.** a (11.8) **17.** c (11.7) **18.** c (11.5) **19.** b (11.5)

Hydrocarbon Derivatives I: Carbon-Heteroatom Single Bonds   Chapter 12

Chapter Overview

Most of the chemical reactions of organic molecules involve the functional groups on the hydrocarbon chains or rings. In this chapter you will consider the properties, chemical and physical, of some hydrocarbon derivatives containing oxygen, sulfur, and nitrogen atoms. Many of these compounds are commonly encountered in naturally occurring substances.

In this chapter you will identify, name, and draw structures for alcohols, phenols, ethers, thiols, and amines. You will compare the physical properties (melting point, boiling point, solubility in water) of these compounds to the hydrocarbons. You will write equations for some of their common reactions.

Practice Exercises

12.1 The additional atoms in hydrocarbon derivatives are called **heteroatoms** (Sec. 12.1). In the table below you will consider some features of five classes of hydrocarbon derivatives. Complete the areas that have been left blank.

Class of hydrocarbon derivative	Heteroatom	Functional group	Generalized formula
ether			
	N		
			R–OH
		–SH	
halogenated hydrocarbon			

12.2 In naming the **halogenated hydrocarbons** (Sec. 12.2) by IUPAC rules, the halogen atoms are treated as substituents on the hydrocarbon molecule. Give the IUPAC names for the molecules shown below:

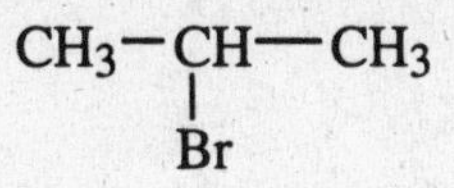

$CH_3-CH-CH_3$
     |
    Br
 	
a.
	$I-\bigcirc-I$
	
	b.
		$\bigcirc\!\!-Cl$ (with two Cl)
		
		c.

$CH_3-CH-CH=CH_2$
     |
    F
 	
d.
	$\triangleright\!\!-Cl$
	
	e.
		$CH_2=CH-CH=CH-CH_2$
		                    |
		                   Br
		
		f.

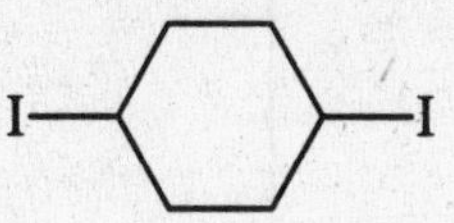

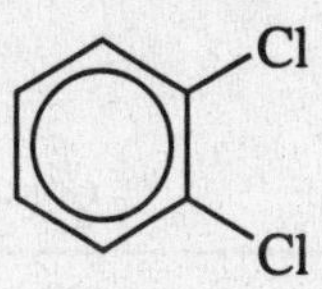

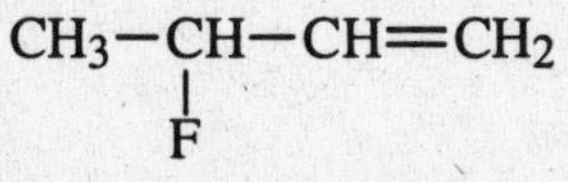

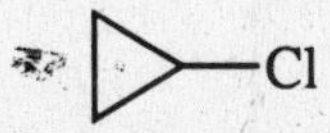

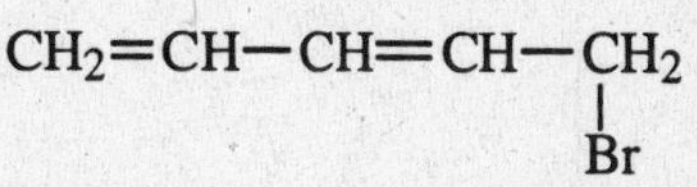

12.3   Draw the structural formulas for the halogenated hydrocarbons named below.

a. 3-iodohexane	b. bromocyclopentane
c. 1-bromo-3-chlorobenzene	d. 1,1-dibromo-1,3-butadiene

12.4   An **alcohol** (Sec. 12.1) is a hydrocarbon derivative in which a **hydroxyl group** (Sec. 12.3) is attached to a saturated carbon atom. A **phenol** (Sec. 12.3) is a compound in which a hydroxyl group is attached to a carbon atom in an aromatic ring system. The names of alcohols and phenols are derived from the hydrocarbons and given the ending *-ol*.

Give the IUPAC name for each of the following compounds:

a. $CH_3-CH_2-CH_2-\overset{\overset{\displaystyle CH_3}{\vert}}{\underset{\underset{\displaystyle CH_3}{\vert}}{C}}-OH$	b. $CH_3-CH_2-CH_2-CH_2-\overset{}{\underset{\underset{\displaystyle OH}{\vert}}{CH}}-\overset{}{\underset{\underset{\displaystyle OH}{\vert}}{CH_2}}$
c. $CH_3-CH_2-\overset{\overset{\displaystyle CH_3}{\vert}}{CH}-\overset{\overset{}{}}{\underset{\underset{\displaystyle CH_3}{\vert}}{CH}}-OH$	d.

12.5   Draw structural formulas for the following alcohols and phenols:

a. 3-hexanol	b. 2,3-dimethyl-1-butanol
c. 2-methyl-2-butanol	d. 3-propylphenol

12.6 Alcohols undergo **dehydration** (Sec. 12.3), an **elimination reaction** (Sec. 12.3) in which a water molecule is removed. When the alcohol is heated in the presence of a sulfuric acid catalyst at 180°C, it produces an alkene. The carbon adjacent to the hydroxyl-bearing carbon loses a hydrogen atom. If there are more than one adjacent carbons, the one with the fewest hydrogen atoms loses a hydrogen atom.

Another important reaction of alcohols is combustion, the reaction with O_2 to produce CO_2 and H_2O.

Complete the following equations by supplying the missing information:

a. $CH_3-CH-CH_3$ + O_2 $\longrightarrow$? + ?
$\quad\quad\quad$ |
$\quad\quad\quad$ OH

b. $CH_3-CH_2-CH_2-CH-CH_3$ $\xrightarrow[H_2SO_4]{180°C}$? + ?
$\quad\quad\quad\quad\quad\quad\quad\quad$ |
$\quad\quad\quad\quad\quad\quad\quad\quad$ OH

c. ? $\xrightarrow[H_2SO_4]{180°C}$ $CH_3-CH_2-CH_2-CH=C-CH_3$ + H_2O
$\quad\quad\quad\quad\quad\quad\quad\quad\quad\quad\quad\quad\quad\quad\quad\quad\quad$ |
$\quad\quad\quad\quad\quad\quad\quad\quad\quad\quad\quad\quad\quad\quad\quad\quad\quad$ CH_3

12.7 **Primary and secondary alcohols** (Sec. 12.4) undergo oxidation in the presence of a mild oxidizing agent to produce compounds that contain a carbon-oxygen double bond. **Tertiary alcohols** (Sec. 12.4) cannot be oxidized in this way since the carbon attached to the hydroxyl group is attached to three other carbons and cannot form a carbon-oxygen double bond.

Alcohols also undergo substitution reactions. An example is the replacement of the hydroxyl group by a halogen atom using PCl_3 or PBr_3.

Complete the following equations by supplying the missing information:

a. $CH_3-CH_2-CH_2-CH_2-CH_2-OH$ $\xrightarrow[\text{agent}]{\text{mild oxidizing}}$?

b. $\xrightarrow[\text{agent}]{\text{mild oxidizing}}$?

c. $CH_3-CH_2-CH_2-CH-CH_3$ $\xrightarrow[\text{agent}]{\text{mild oxidizing}}$?
$\quad\quad\quad\quad\quad\quad\quad\quad$ |
$\quad\quad\quad\quad\quad\quad\quad\quad$ OH

d. $CH_3-CH-CH_2-CH_3$ $\xrightarrow{PCl_3}$?
$\quad\quad\quad$ |
$\quad\quad\quad$ OH

12.8   Phenols can act as weak acids in the presence of bases. Complete the following reaction showing this property.

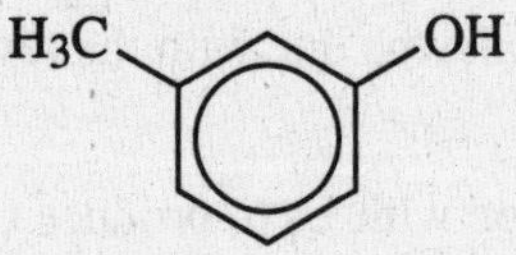

$$H_3C - \bigotimes - OH \quad + \quad NaOH(aq) \quad \longrightarrow \quad ? \quad + \quad ?$$

12.9   **Ethers** (Sec. 12.5) are compounds with two hydrocarbon groups attached to an oxygen atom. According to the IUPAC system, ethers are named as substituted hydrocarbons. The smaller hydrocarbon and the oxygen atom are called an **alkoxy group** (Sec. 12.5) and considered as a substituent on the larger hydrocarbon.

Write the IUPAC name for each of the following compounds:

$CH_3-CH_2-CH_2-CH_2-O-CH_3$	$\bigcirc - O-CH_2-CH_3$
a.	b.

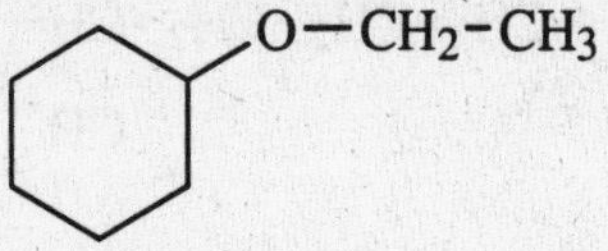

12.10   Draw the condensed structural formula and the common name for each of the following ethers:

2-ethoxypropane	propoxycyclopentane
Common name:	Common name:

12.11   Some physical properties of organic compounds having oxygen-containing functional groups are determined by a molecule's ability to form hydrogen bonds. If a molecule can form hydrogen bonds with a like molecule (as alcohols can), the substance will have a higher boiling point than would be expected for its molecular mass. If a substance can form hydrogen bonds with water (as both alcohols and ethers can), its solubility in water is greater than a hydrocarbon of similar molecular mass.

Indicate which of the following pairs of compounds would have the higher boiling point and the greater solubility in water.

Compounds	Higher boiling point	Greater solubility in water
1-propanol and propane		
diethyl ether and 2-methyl-2-propanol		
phenol and toluene		

12.12 Sulfur analogs of alcohols are called **thiols** (Sec. 12.6). These compounds contain a **sulfhydryl group** (Sec. 12.6) and are named by adding -*thiol* to the name of the parent alkane.

Give the IUPAC name for the following thiols:

$CH_3-CH_2-CH_2-CH_2-CH_2-SH$ a.	 b.

12.13 Write the structural formulas for the following thiols:

    a. 2-methyl-1-butanethiol	 b. 2-ethylcyclobutanethiol

12.14 Use this identification exercise to review your knowledge of functional groups to this point. Using structures A through I, give the best match for the terms below:

$HO-CH_2-CH_2-OH$ A.	$CH_3-O-CH_2-CH_3$ B.	$CH_3-CH_2-CH-CH_3$ $\quad\quad\quad\quad\;\; \mid$ $\quad\quad\quad\quad\;\; Cl$ C.
$\quad\quad\quad\quad CH_3$ $\quad\quad\quad\quad \mid$ CH_3-CH_2-C-OH $\quad\quad\quad\quad \mid$ $\quad\quad\quad\quad CH_3$ D.	—OH E.	$CH_3-CH-CH_3$ $\quad\quad\;\; \mid$ $\quad\quad\;\; OH$ F.
$CH_3-CH_2-CH{=}CH_2$ G.	CH_3-CH_2-SH H.	$CH_3-S-S-CH_3$ I.

a. 3° alcohol _____

b. glycol _____

c. halogenated hydrocarbon____

d. ether _____

e. phenol _____

f. alkene _____

g. thiol _____

h. disulfide _____

i. 2° alcohol _____

Give the IUPAC names for A through H.

j. structure A _____

k. structure B _____

l. structure C _____

m. structure D_____

n. structure E_____

o. structure F_____

p. structure G_____

q. structure H_____

12.15 **Amines** (Sec. 12.7) are organic derivatives of ammonia in which one or more hydrogens have been replaced by an alkyl, cycloalkyl, or aryl group. Both common and IUPAC names are used for amines.

Complete the following table. Follow the rules for naming given in Section 12.7 of your textbook.

Structural formula	IUPAC name	Common name
a. $CH_3-CH_2-CH_2-NH_2$		
b. $CH_3-CH_2-CH_2-N\begin{smallmatrix}CH_3\\CH_3\end{smallmatrix}$		
c. $HN-CH_2-CH_2-CH_3$ $\quad CH_2-CH_3$		
d. ⟨benzene ring⟩$-NH_2$		
e. ⟨benzene ring⟩$-NH_2$ with CH_2-CH_3		

12.16 Amines are classified as primary, secondary, or tertiary (Section 12.7) on the basis of the number of alkyl groups attached to the nitrogen atom. Classify each of the amines in the practice exercise above as a primary, seconday, or tertiary amine.

a.

b.

c.

d.

e.

12.17 Since amines are bases, their reaction with an acid produces an **amine salt** (Sec, 12.8). The amine may be regenerated by reaction of the salt with a base. Complete the following equations.

a. 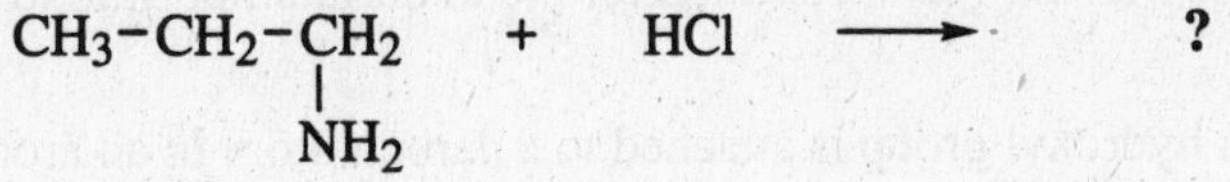 $CH_3-CH_2-CH_2$ + HCl $\longrightarrow$?
       |
      NH_2

b. 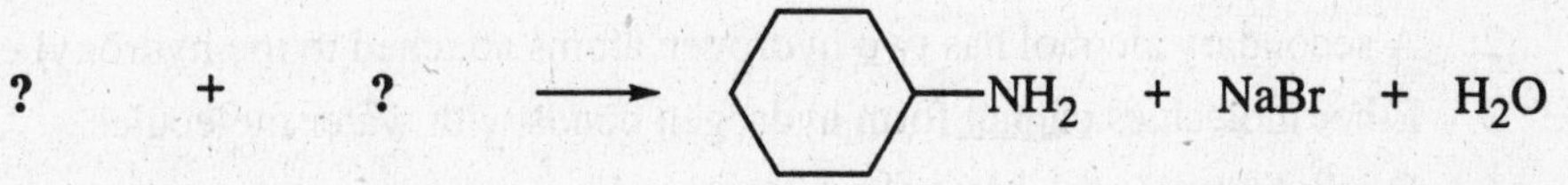 ? + ? $\longrightarrow$ ⬡$-NH_2$ + NaBr + H_2O

12.18 Amine salts are named in the same way as other ionic compounds are: the positive ion, the **substituted ammonium ion** (Sec. 12.8), is named first, followed by the name of the negative ion. Give the name of each of the amine salts in the practice exercise above.

a.

b.

12.19 The identification exercise below is a review of the common structures presented in Sections 12.7 through 12.9 in your textbook. Write letters A through F after any terms to which they apply. Some answers apply to more than one structure.

$CH_3-CH_2-CH_2-NH_2$ A.	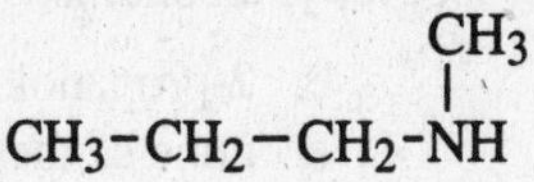 CH_3 | $CH_3-CH_2-CH_2-NH$ B.	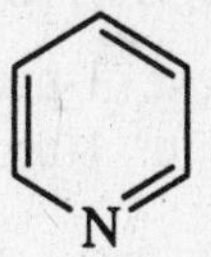 C.
$CH_3-CH_2-\overset{+}{N}H_3$ Cl^- D.	 NH_2 ⬡ E.	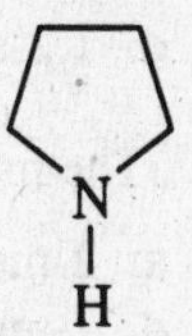 N | H F.

a. heterocyclic amine _____

b. pyridine _____

c. aniline _____

d. amine salt _____

e. primary amine _____

f. secondary amine _____

Self-Test

True-false: Indicate whether the following statements are true or false. If the statement is false, give the word or phrase that may be substituted for the underlined portion to make the statement true.

1. A compound in which a hydroxyl group is attached to a carbon atom in an aromatic ring system is called a <u>thiol</u>.

2. A secondary alcohol has <u>two</u> hydrogen atoms attached to the hydroxyl carbon.

3. Ether molecules <u>cannot</u> form hydrogen bonds with water molecules.

4. Diethyl ether and <u>1-butanol</u> are structural isomers.

5. 2-Butanol is an example of a <u>tertiary</u> alcohol.

6. Alcohols are <u>more soluble</u> in water than alkanes of similar molecular mass.

7. Purine is an example of a <u>heterocyclic</u> organic compound.

8. Storage of ethers can be hazardous if unstable <u>peroxides</u> form.

9. Nitrogen forms <u>two covalent bonds</u> to complete its octet of electrons.

10. Dimethylamine is a <u>primary amine</u>.

11. The IUPAC name for an amine having two methyl groups and one ethyl group attached to a nitrogen atom is <u>ethyldimethylamine</u>.

12. A benzene ring with an attached amino group is called <u>aniline</u>.

13. Thiols are noted for their <u>pleasant odors</u>.

14. Reaction of an amine with <u>a base</u> produces an amine salt.

Multiple choice:

15. The IUPAC name for isopropyl alcohol is:

    a. 1-propyl alcohol   b. 2-propanol   c. 2-propyl alcohol
    d. 1-propanol   e. none of these

16. The common name for 1,2-ethanediol is:

    a. ethylene glycol   b. glycerol   c. diethyl ether
    d. 1,2-ethyl alcohol   e. glycerin

17. The main product of the dehydration of 2-methyl-2-butanol would be:

    a. 2-methyl-2-butene   b. 2-methyl-1-butene   c. 1-methyl-2-butene
    d. 1-methyl-1-butene   e. none of these

18. Which of the following is a correct name according to IUPAC rules?

    a. 2-ethyl-1-butene-2-ol   b. 3-propane-2-ol   c. 3,4-butanediol
    d. 1,2-dimethylphenol   e. none of these

19. An amine can be formed from its amine salt by treating the salt with:

    a. NaOH   b. an alcohol   c. H_2SO_4
    d. a carboxylic acid   e. none of these

20. Which of the following is **not** a heterocyclic amine derivative?

    a. caffeine   b. codeine   c. heme
    d. aniline   e. heroin

12.9   a.   1-methoxybutane;
       b.   ethoxycyclohexane

12.10

2-ethoxypropane	propoxycyclopentane
CH_3	
$CH_3-CH_2-O-\overset{\|}{CH}-CH_3$	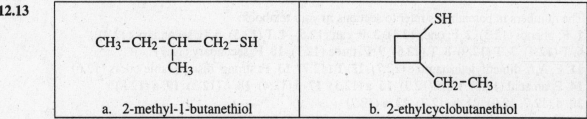
ethyl isopropyl ether	cyclopentyl propyl ether

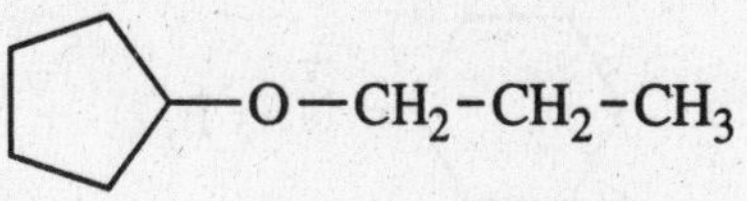

12.11

Compounds	Higher boiling point	Greater solubility in water
1-propanol and propane	1-propanol	1-propanol
diethyl ether and 2-methyl-2-propanol	2-methyl-2-propanol	2-methyl-2-propanol
phenol and toluene	phenol	phenol

12.12   a.   1-pentanethiol;
        b.   cyclohexanethiol

12.13

$CH_3-CH_2-\overset{\|}{CH}-CH_2-SH$ $\quad CH_3$	(cyclobutane with SH and CH_2-CH_3)
a. 2-methyl-1-butanethiol	b. 2-ethylcyclobutanethiol

12.14   a. D; b. A; c. C; d. B; e. E; f. G; g. H; h. I; i. F
       j.   1,2-ethanediol
       k.   methoxyethane
       l.   2-chlorobutane
       m.   2-methyl-2-butanol
       n.   phenol
       o.   2-propanol
       p.   1-butene
       q.   ethanethiol

12.15   a.   1-propanamine; propylamine
       b.   *N,N*-dimethyl-1-propanamine; dimethylpropylamine
       c.   *N*-ethyl-1-propanamine; ethylpropylamine
       d.   aniline; aniline
       e.   2-ethylaniline or *o*-ethylaniline

12.16   a.   primary
       b.   tertiary
       c.   secondary
       d.   primary
       e.   primary

12.17 a.

$$CH_3-CH_2-CH_2 \quad + \quad HCl \quad \longrightarrow \quad CH_3-CH_2-CH_2$$
$$\overset{|}{NH_2} \qquad\qquad\qquad\qquad\qquad\qquad \overset{|}{\overset{+}{N}H_3} \ Cl^-$$

b.

⬡—$\overset{+}{N}H_3$ Br^- + NaOH ⟶ ⬡—NH_2 + NaBr + H_2O

12.18 a. propylammonium chloride
b. cyclohexylammonium bromide

12.19 a. C. and F.
b. C.
c. E.
d. D.
e. A. and E.
f. F. and B.

Answers to Self-Test

The numbers in parentheses refer to sections in your textbook:
1. F; phenol (12.3) **2.** F; one (12.4) **3.** F; can (12.5) **4.** T (12.5) **5.** F; secondary (12.4)
6. T (12.4) **7.** T (12.9) **8.** T (12.5) **9.** F; three (12.7) **10.** F; secondary (12.7)
11. F; N,N-dimethylethanamine (12.7) **12.** T (12.7) **13.** F; strong, disagreeable odors (12.6)
14. F; an acid (12.8) **15.** b (12.3) **16.** a (12.3) **17.** a (12.4) **18.** e (12.3) **19.** a (12.8)
20. d (12.7, 12.9) **21.** b (12.7) **22.** a (12.7)

Hydrocarbon Derivatives II:
Carbon-Oxygen Double Bonds
Chapter 13

Chapter Overview

The carbonyl functional group is very commonly found in nature. It contains an oxygen atom joined to a carbon atom by a double bond. Aldehydes, ketones, carboxylic acids, esters, and amides all have functional groups that contain the carbonyl group.

In this chapter you will learn to recognize, name, and write structural formulas for compounds that contain the carbonyl group. You will compare the physical properties of these compounds with one another and with the hydrocarbon derivatives you studied in the last chapter. You will learn to write equations for some of the typical reactions involving these classes of compounds.

Practice Exercises

13.1   Five major classes of organic compounds are linked by one feature: the presence of a **carbonyl group** (Sec. 13.1), a carbon atom and an oxygen atom joined by a double bond.

Complete the table below to practice recognizing the structures and functional groups of carbonyl-containing compounds.

General name of class	Functional group	Simplest example	General structure
aldehydes			
	$\overset{O}{\underset{\|}{\overset{\|}{—C—OH}}}$		
		$CH_3—\overset{O}{\overset{\|}{\underset{\|}{C}}}—CH_3$	
			$R—\overset{O}{\overset{\|}{\underset{\|}{C}}}—NH_2$
esters			

13.2   **Aldehydes** (Sec. 13.2) are compounds in which the carbonyl carbon is bonded to at least one hydrogen. **Ketones** (Sec. 13.4) are compounds in which the carbonyl carbon is bonded to two other carbons.

Classify each of the following structural formulas as an aldehyde, a ketone, or neither.

CH₃–CH₂–CH₂–O–CH₃ a.	$$CH_3-\underset{\underset{CH_3}{\mid}}{CH}-\overset{\overset{O}{\parallel}}{C}-H$$ b.	benzaldehyde structure c.
H₃C– (cyclohexanone) =O d.	$$CH_3-\underset{\underset{C_6H_5}{\mid}}{CH}-\overset{\overset{O}{\parallel}}{C}-CH_3$$ e.	$$H-\overset{\overset{O}{\parallel}}{C}-CH_2-\underset{\underset{CH_3}{\mid}}{\overset{\overset{CH_3}{\mid}}{C}}-CH_3$$ f.

13.3   In naming aldehydes, change the *-e* ending of the hydrocarbon to *-al*, and name any substituents, starting the counting from the carbonyl carbon. No number is specified for the carbonyl group.

Give the IUPAC name for the three aldehydes in the Practice Exercise above (13.2).

13.4   Draw the structural formula for each of the following aldehydes.

a. 4-chloro-3-methylbutanal	b. 5-bromo-3-methylpentanal

13.5   Three common reactions of aldehydes are reduction, oxidation and **hemiacetal** and **acetal** (Sec. 13.3) formation. Reduction of an aldehyde with various reducing agents (such as hydrogen gas with a Ni catalyst) produces a primary alcohol. Aldehydes are readily oxidized to carboxylic acids. The Tollens' and Benedict's tests for aldehydes are based on this ease of oxidation.

The carbonyl group can undergo addition reactions. The addition of one molecule of an alcohol to one molecule of an aldehyde produces a hemiacetal. The addition of two molecules of alcohol to one molecule of an aldehyde produces an acetal.

Complete the equations below showing reactions of aldehydes:

a. $CH_3-CH_2-CH_2-\overset{\overset{\displaystyle O}{\|}}{C}-H$ + H_2 $\xrightarrow{Ni}$ **?**

b.   **?**   +   **?** $\xrightarrow{Ni}$ $CH_3-CH_2-CH_2-OH$

c. $CH_3-CH_2-CH_2-\overset{\overset{\displaystyle O}{\|}}{C}-H$ $\xrightarrow{\overset{\text{mild}}{\text{oxidizing agent}}}$ **?**

d. $CH_3-CH_2-\overset{\overset{\displaystyle O}{\|}}{C}-H$ + 1 CH_3-OH $\overset{H^+}{\rightleftharpoons}$ **?**

e.   **?** $\xrightarrow{\overset{\text{mild}}{\text{oxidizing agent}}}$ $CH_3-CH_2-\overset{\overset{\displaystyle O}{\|}}{C}-OH$

f.   **?**   +   **?** $\overset{H^+}{\rightleftharpoons}$ $CH_3-CH_2-\underset{\underset{\displaystyle O-CH_2-CH_3}{|}}{\overset{\overset{\displaystyle O-CH_2-CH_3}{|}}{C}}-H$

13.6   In naming ketones, change the -*e* ending to -*one*. Give the carbonyl group the lowest possible number on the chain, and then name the substituents. The numbered position of the carbonyl group is included in the name.

Give the IUPAC names for the following ketones.

$CH_3-\overset{\overset{\displaystyle O}{\|}}{C}-\overset{\overset{\displaystyle Cl}{\|}}{C}H-\underset{\underset{\displaystyle CH_3}{\|}}{C}H-CH_3$	$CH_3-\underset{\underset{\displaystyle CH_3}{\|}}{C}H-\overset{\overset{\displaystyle O}{\|}}{C}-CH_3$	$H_3C-\hexagon=O$
a.	b.	c.

13.7   Draw the structural formula for each of the following ketones.

a. 4-methyl-3-hexanone	b. 3-ethyl-2-methylcyclopentanone

13.8   Ketones are easily reduced to secondary alcohols with hydrogen gas in the presence of a catalyst. Ketones are not oxidized by mild oxidizing agents, as aldehydes are, since the carbonyl group in ketones is not bonded to a hydrogen atom.

Ketones form **hemiketals** and **ketals** (Sec. 13.5) in a manner similar to the aldehyde formation of hemiacetals and acetals: the addition of one molecule of an alcohol to one molecule of a ketone forms a hemiketal, and the addition of two molecules of alcohol to one molecule of a ketone forms a ketal.

Complete the following equations showing reactions of ketones:

a. $CH_3-\overset{\overset{\displaystyle O}{\|}}{C}-CH_2-CH_3$ + H_2 $\xrightarrow{\text{Ni}}$?

b. ? + ? $\xrightarrow{\text{Ni}}$ $CH_3-\overset{\overset{\displaystyle OH}{|}}{CH}-CH_3$

c. $CH_3-\overset{\overset{\displaystyle O}{\|}}{C}-CH_3$ + 1 CH_3-OH $\underset{}{\overset{H^+}{\rightleftharpoons}}$?

d. $CH_3-CH_2-\overset{\overset{\displaystyle O}{\|}}{C}-CH_3$ $\xrightarrow[\text{oxidizing agent}]{\text{mild}}$?

e. ? + ? $\overset{H^+}{\rightleftharpoons}$ $CH_3-CH_2-\overset{\overset{\displaystyle O-CH_2-CH_3}{|}}{\underset{\underset{\displaystyle O-CH_2-CH_3}{|}}{C}}-CH_3$

13.9 Hydrogen bonding cannot occur between the molecules of an aldehyde or between the molecules of a ketone. However, dipole-dipole attractions occur between molecules.

Determine which one of each pair of compounds would have a higher boiling point, and explain your choice.

Compounds	Higher boiling point	Explanation
pentanal and hexane		
2-hexanone and octane		
pentanal and 1-pentanol		

13.10 **Carboxylic acids** (Sec. 13.6) contain the **carboxyl group** (Sec. 13.6), a carbonyl group with a hydroxyl group bonded to the carbon atom. The naming of carboxylic acids is similar to that of aldehydes, with the ending *–oic acid.*

Give the IUPAC name for each of the following carboxylic acids.

$CH_3-\overset{\overset{\displaystyle O}{\|}}{\underset{\underset{\displaystyle CH_3}{|}}{CH}}-C-OH$	$I-\bigcirc-\overset{\overset{\displaystyle O}{\|}}{C}-OH$	$CH_3-\overset{\overset{\displaystyle CH_3}{|}}{CH}-\overset{\overset{\displaystyle}{}}{\underset{\underset{\displaystyle Cl}{|}}{CH}}-CH_2-\overset{\overset{\displaystyle O}{\|}}{C}-OH$
a.	b.	c.

13.11 Draw the structural formulas for the following carboxylic acids:

a. 3-chloro-2-methylpropanoic acid	b. 2,2-dibromobutanoic acid

13.12 Common names are used for many of the carboxylic acids and **dicarboxylic acids** (Sec. 13.6). Complete the following table to compare the common name and the IUPAC name for some acids.

IUPAC name	Common name	Structural formula
ethanoic acid		
methanoic acid		
butanedioic acid		

13.13 Carboxylic acids have very high boiling points because two molecules can form two hydrogen bonds with one another, from each of the double-bonded oxygens to the hydrogens of the –OH groups. Carboxylic acids also form hydrogen bonds with water, and so are somewhat water soluble, especially those with short hydrocarbon chains.

Draw diagrams of the hydrogen bonding between the following molecules. Use structural formulas and show hydrogen bonds as dotted lines.

a. two molecules of propanoic acid	b. one molecule of propanoic acid and water

13.14 Carboxylic acids react with strong bases to produce water and a **carboxylic acid salt** (Sec. 13.7). The negative ion of the salt is called a **carboxylate ion** (Sec. 13.7).

$$R-\overset{\overset{\displaystyle O}{\|}}{C}-OH \ + \ NaOH \ \longrightarrow \ R-\overset{\overset{\displaystyle O}{\|}}{C}-O^- \ Na^+ \ + \ H_2O$$

Complete the following table of carboxylic acis and their sodium salts.

Name of acid	Structural formula of carboxylic acid salt	IUPAC name of carboxylic acid salt
propanoic acid		
	$CH_3-CH_2-CH_2-\overset{\overset{\displaystyle O}{\|}}{C}-O^- \ Na^+$	
benzoic acid		

13.15 A carboxylic acid salt can be converted to the carboxylic acid by reacting the salt with a strong acid such as HCl or H_2SO_4.

Write an equation for the reaction of sodium propanoate with hydrochloric acid to form the carboxylic acid.

13.16 **Esters** (Sec. 13.8) are carboxylic acid derivatives in which the –OH group has been replaced with an –OR group. They are usually produced by the reaction of a carboxylic acid with an alcohol, with a strong acid catalyst. This reaction is called esterification.

$$R-\overset{\overset{\textstyle O}{\|}}{C}-OH \quad + \quad R'-OH \quad \underset{}{\overset{H^+}{\rightleftharpoons}} \quad R-\overset{\overset{\textstyle O}{\|}}{C}-O-R'$$

Complete the following table showing the "parent" acid and "parent" alcohol, and the ester that is formed in their reaction. In ester names, the alcohol portion is named first followed by the acid part with an *–ate* ending.

Acid and alcohol	Structure of ester formed	IUPAC name of ester
ethanoic acid and methanol		
	$CH_3-CH_2-CH_2-\overset{\overset{\textstyle O}{\|}}{C}-O-CH_3$	
		ethyl benzoate

13.17 Esters are often known by common names, which are based on the common names of the acid parts of the ester. Complete the following table comparing the IUPAC and common names of some esters.

IUPAC name	Common name
ethyl methanoate	
	propyl butyrate
methyl ethanoate	

13.18 Esters undergo a **hydrolysis reaction** (Sec. 13.9) with an acid catalyst to produce the acid and alcohol from which they were formed. In the process, a molecule of water is split into –H and –OH and added to the two componenets of the reactant.

Write the equation for the hydrolysis reaction of ethyl butanoate with a strong acid catalyst.

13.19 An **amide** (Sec. 13.10) is a carboxylic acid derivative in which the carboxyl –OH group is replaced by an amino or substituted amino group. Amides undergo hydrolysis when heated with a catalyst, producing the "parent" carboxylic acid and "parent" amine (or ammonia). Complete the following hydrolysis reactions:

a.
$$CH_3-CH_2-CH_2-\overset{\overset{\displaystyle O}{\|}}{C}-NH_2 \quad + \quad H_2O \quad \xrightarrow{\text{heat}} \quad ?$$

b.
$$CH_3-\underset{\underset{\displaystyle Cl}{|}}{CH}-\overset{\overset{\displaystyle O}{\|}}{C}-\overset{\overset{\displaystyle CH_3}{|}}{N}-CH_2-CH_3 \quad + \quad H_2O \quad \xrightarrow{\text{heat}} \quad ?$$

c.
$$\underset{Br}{\text{[benzene ring]}}-\overset{\overset{\displaystyle O}{\|}}{C}-\overset{\overset{\displaystyle H}{|}}{N}-CH_3 \quad + \quad H_2O \quad \xrightarrow{\text{heat}} \quad ?$$

13.20 Amides can be classified as primary, secondary, or tertiary, based on the number of carbon atoms bonded to the nitrogen atom. Primary amides have an unsubstituted amino group, while one or two hydrogens are replaced with –R groups for secondary and tertiary amides.

Classify each of the amides below as primary, secondary, or tertiary, and give the IUPAC name.

Structural formula	Primary, secondary, or tertiary amide	IUPAC name of amide
$CH_3-\overset{\overset{\displaystyle O}{\|}}{C}-NH-CH_3$		
$CH_3-CH_2-CH_2-\overset{\overset{\displaystyle O}{\|}}{C}-NH_2$		
$\text{[benzene ring]}-\overset{\overset{\displaystyle O}{\|}}{C}-\overset{\overset{\displaystyle CH_3}{|}}{N}-CH_3$		

13.21 Draw structural formulas for the following substituted amides.

a. *N*-ethyl-*N*-propylpentamide	b. *N,N*-dimethylbenzamide

13.22 Amides undergo hydrolysis in a manner similar to that of esters. When heated with a catalyst, the amide is split into the "parent" carboxylic acid and the "parent" amine (or ammonia, for a primary amide). A molecule of water is split and added to the two reactant components.

$$\underset{R-\overset{\displaystyle O}{\overset{\|}{C}}-NH_2}{} \quad + \quad H-OH \quad \overset{heat}{\longrightarrow} \quad \underset{R-\overset{\displaystyle O}{\overset{\|}{C}}-OH}{} \quad + \quad H-NH_2$$

Complete the following equations for the hydrolysis of each of these amides:

a. 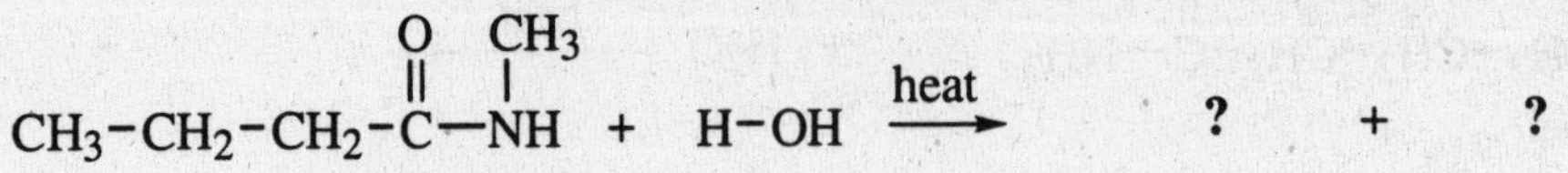 $CH_3-CH_2-CH_2-\overset{\displaystyle O}{\overset{\|}{C}}-\overset{\displaystyle CH_3}{\overset{|}{NH}}$ +   H-OH $\overset{heat}{\longrightarrow}$ **?**   +   **?**

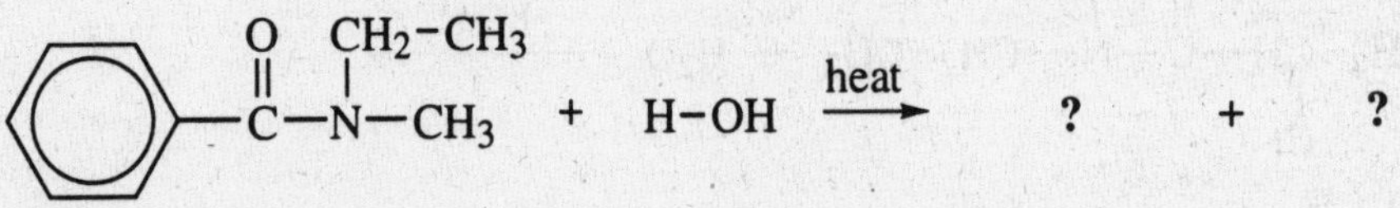

b. $\bigcirc\!\!\!-\overset{\displaystyle O}{\overset{\|}{C}}-\overset{\displaystyle CH_2-CH_3}{\overset{|}{N}}-CH_3$ +   H-OH $\overset{heat}{\longrightarrow}$ **?**   +   **?**

c. $CH_3-CH_2-\overset{\displaystyle O}{\overset{\|}{C}}-NH_2$ +   H-OH $\overset{heat}{\longrightarrow}$ **?**   +   **?**

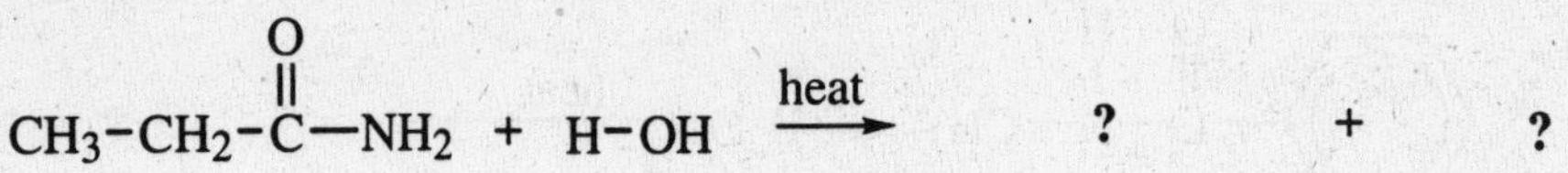

13.23   Use this identification exercise to review your knowledge of the structures in this chapter. Using letters A. through I., give the best choice for the list of terms beneath the table.

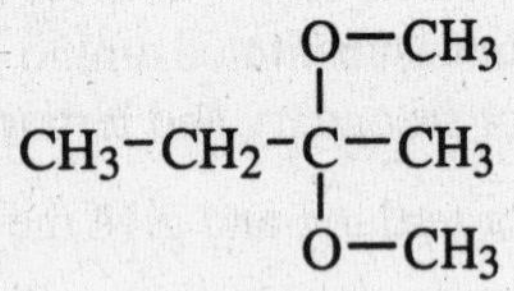

$CH_3-CH_2-\overset{\displaystyle O-CH_3}{\underset{\displaystyle O-CH_3}{\overset{|}{\underset{|}{C}}}}-CH_3$ A.	$CH_3-\overset{\displaystyle O}{\overset{\|}{C}}-CH_3$ B.	$CH_3-CH_2-\overset{\displaystyle O-CH_3}{\underset{\displaystyle O-CH_3}{\overset{|}{\underset{|}{C}}}}-H$ C.
$CH_3-\overset{\displaystyle O}{\overset{\|}{C}}-\overset{\displaystyle CH_3}{\overset{|}{N}}-CH_3$ D.	$CH_3-\overset{\displaystyle O}{\overset{\|}{C}}-O^- Na^+$ E.	$CH_3-\overset{\displaystyle O}{\overset{\|}{C}}-OH$ F.
$CH_3-CH_2-\overset{\displaystyle OH}{\underset{\displaystyle O-CH_3}{\overset{|}{\underset{|}{C}}}}-H$ G.	$CH_3-\overset{\displaystyle O}{\overset{\|}{C}}-O-CH_3$ H.	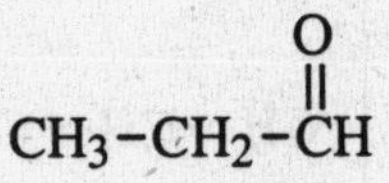 $CH_3-CH_2-\overset{\displaystyle O}{\overset{\|}{CH}}$ I.

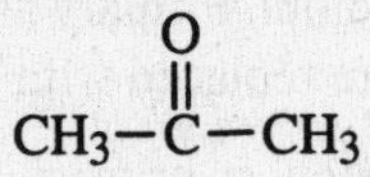

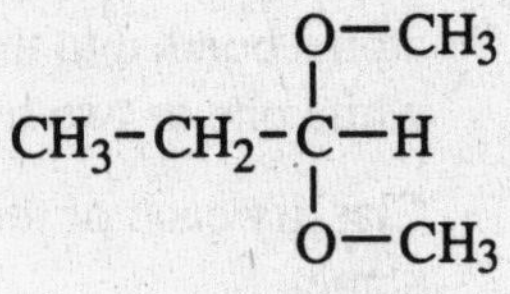

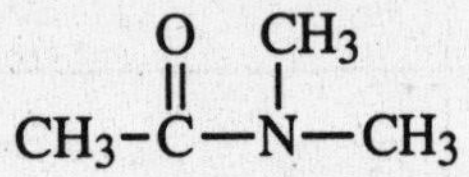

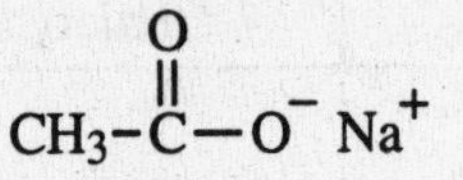

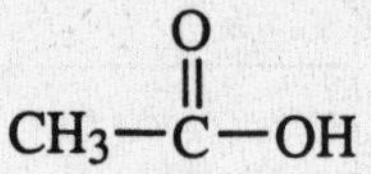

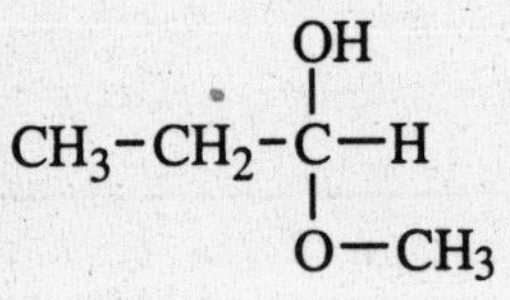

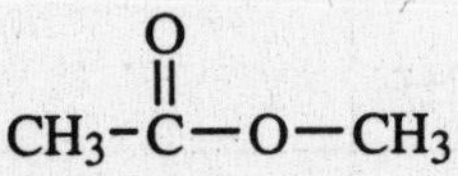

a.   acetal   _____

b.   aldehyde   _____

c.   amide   _____

d.   carboxylate salt   _____

e.   carboxylic acid   _____

f.   ester   _____

g.   ketal   _____

h.   hemiacetal   _____

i.   ketone   _____

Self-Test

True-false: Indicate whether the following statements are true or false. If the statement is false, give the word or phrase that may be substituted for the underlined portion to make the statement true.

1. The simplest ketone contains <u>two</u> carbon atoms.

2. The simplest aldehyde has the common name <u>formaldehyde</u>.

3. The oxidation of 2-butanol produces a <u>ketone</u>.

4. Aldehydes and ketones have <u>higher</u> boiling points than the corresponding alcohols.

5. A positive Tollens' test indicates that <u>an aldehyde</u> is present.

6. A ketone can be <u>oxidized</u> by hydrogen gas in the presence of a Ni catalyst.

7. An acetal molecule is the product of the addition of <u>two molecules</u> of an alcohol to one molecule of an aldehyde.

8. <u>Acetic acid</u> is the simplest carboxylic acid.

9. A carboxylic acid with six carbons in a straight chain is named <u>1-hexanoic acid</u>.

10. A carboxylate salt is formed by the reaction of a carboxylic acid and <u>a strong base.</u>

11. Because of their extensive hydrogen bonding, carboxylic acids have <u>low boiling points</u>.

12. The reaction between sodium hydroxide and benzoic acid would produce <u>sodium benzoate</u>.

13. The pleasant fragrances of many flowers and fruits are produced by mixtures of <u>carboxylic acids</u>.

14. The boiling points of esters are <u>lower</u> than those of alcohols and acids with comparable molecular mass.

15. A polyester is <u>an addition polymer</u> with ester linkages.

16. The various types of nylon are synthesized by the reactions of <u>diamines with dicarboxylic acids</u>.

17. Kevlar is a very tough polyamide that owes its strength to the presence of <u>aromatic rings</u> in its "backbone."

18. Secondary amides have <u>two</u> hydrogens attached to nitrogen.

Multiple choice:

19. The IUPAC name for the compound $CH_3-CH_2-CH_2-CH_2-CHO$ is:

    a. 1-pentanone   b. 1-pentyl ketone   c. 1-pentanal
    d. pentanal   e. none of these

20. A common name for propanone is:

    a. acetone   b. methyl phenyl ketone   c. ethyl methyl ketone
    d. diethyl ketone   e. none of these

21. A hemiketal molecule is formed by the reaction between:

    a. two ketone molecules
    b. two aldehyde molecules
    c. a ketone molecule and an alcohol molecule
    d. a ketone molecule and an aldehyde molecule
    e. none of these

22. How many ketones are isomeric with butanal?

    a. one   b. two   c. three   d. four   e. none of these

23. The reaction between methanol and ethanoic acid will produce:

    a. methyl acetate   b. ethyl formate   c. methyl butyrate
    d. ethyl ethanoate   e. none of these

24. Rank these three types of compounds (of comparable molecular mass) – alcohol, carboxylic acid, ester of carboxylic acid – in order of boiling point, highest to lowest:

    a. alcohol, acid, ester   b. ester, acid, alcohol   c. acid, alcohol, ester
    d. ester, alcohol, acid   e. acid, ester, alcohol

25. One of the products of the hydrolysis of an ester is:

    a. an aldehyde   b. a ketone   c. an alcohol
    d. an ether   e. none of these

26. An example of a dicarboxylic acid is:

    a. succinic acid   b. butyric acid   c. acetic acid
    d. lactic acid   e. none of these

27. The hydrolysis of methyl butanoate in the presence of strong acid would yield:

    a. methanoic acid and butyric acid   b. methanol and butanoic acid
    c. methanoic acid and 1-butanol   d. sodium butanoate and methanol
    e. none of these

28. The hydrolysis of *N*-methylpropanamide yields:

    a. *N*-propanamine and methanol   b. propylamine and acetic acid
    c. methylamine and propanoic acid   d. acetic acid and methylamine
    e. none of these

Answers to Practice Exercises

13.1

General name of class	Functional group	Simplest example	General structure
aldehydes	$\underset{\underset{C-C-H}{}}{\overset{O}{\parallel}}$	$\underset{\underset{H-C-H}{}}{\overset{O}{\parallel}}$	$\underset{\underset{R-C-H}{}}{\overset{O}{\parallel}}$
carboxylic acids	$\underset{\underset{-C-OH}{}}{\overset{O}{\parallel}}$	$\underset{\underset{CH_3-C-OH}{}}{\overset{O}{\parallel}}$	$\underset{\underset{R-C-OH}{}}{\overset{O}{\parallel}}$
ketones	$\underset{\underset{C-C-C}{}}{\overset{O}{\parallel}}$	$\underset{\underset{CH_3-C-CH_3}{}}{\overset{O}{\parallel}}$	$\underset{\underset{R-C-R'}{}}{\overset{O}{\parallel}}$
amides	$\underset{\underset{-C-NH_2}{}}{\overset{O}{\parallel}}$	$\underset{\underset{H-C-NH_2}{}}{\overset{O}{\parallel}}$	$\underset{\underset{R-C-NH_2}{}}{\overset{O}{\parallel}}$
esters	$\underset{\underset{-C-O-C}{}}{\overset{O}{\parallel}}$	$\underset{\underset{H-C-O-CH_3}{}}{\overset{O}{\parallel}}$	$\underset{\underset{R-C-O-R'}{}}{\overset{O}{\parallel}}$

13.2 a. neither (an ether); b. aldehyde; c. aldehyde; d. ketone; e. ketone; f. aldehyde;

13.3 b. 2-methylpropanal; c. benzaldehyde; f. 3,3-dimethylbutanal

13.4

a. 4-chloro-3-methylbutanal	b. 5-bromo-3-methylpentanal
$\underset{\underset{Cl}{\mid}}{CH_2}-\underset{\underset{CH_3}{\mid}}{CH}-CH_2-\overset{\overset{O}{\|}}{C}-H$	$\underset{\underset{Br}{\mid}}{CH_2}-CH_2-\underset{\underset{CH_3}{\mid}}{CH}-CH_2-\overset{\overset{O}{\|}}{C}-H$

13.5

a. $CH_3-CH_2-CH_2-\overset{\overset{O}{\|}}{C}-H + H_2 \xrightarrow{Ni} CH_3-CH_2-CH_2-CH_2-OH$

b. $CH_3-CH_2-\overset{\overset{O}{\|}}{C}-H + H_2 \xrightarrow{Ni} CH_3-CH_2-CH_2-OH$

c. $CH_3-CH_2-CH_2-\overset{\overset{O}{\|}}{C}-H \xrightarrow{\text{mild oxidizing agent}} CH_3-CH_2-CH_2-\overset{\overset{O}{\|}}{C}-OH$

d. $CH_3-CH_2-\overset{\overset{O}{\|}}{C}-H + 1\ CH_3-OH \overset{H^+}{\rightleftharpoons} CH_3-CH_2-\underset{\underset{O-CH_3}{\mid}}{\overset{\overset{OH}{\mid}}{C}}-H$

e. $CH_3-CH_2-\overset{\overset{O}{\|}}{C}-H \xrightarrow{\text{mild oxidizing agent}} CH_3-CH_2-\overset{\overset{O}{\|}}{C}-OH$

f. $CH_3-CH_2-\overset{\overset{O}{\|}}{C}-H + 2\ CH_3-CH_2-OH \overset{H^+}{\rightleftharpoons} CH_3-CH_2-\underset{\underset{O-CH_2-CH_3}{\mid}}{\overset{\overset{O-CH_2-CH_3}{\mid}}{C}}-H$

13.6 a. 3-chloro-4-methyl-2-pentanone;
b. 3-methyl-2-butanone;
c. 4-methylcyclohexanone

13.7

a. 4-methyl-3-hexanone	b. 3-ethyl-2-methylcyclopentanone
$CH_3-CH_2-\underset{\underset{CH_3}{\mid}}{CH}-\overset{\overset{O}{\|}}{C}-CH_2-CH_3$	(cyclopentanone ring with =O at top, CH_3 at position 2, CH_2-CH_3 at position 3)

13.8 a. $CH_3-\overset{\displaystyle O}{\overset{\|}{C}}-CH_2-CH_3$ + H_2 $\xrightarrow{Ni}$ $CH_3-\overset{\displaystyle OH}{\overset{|}{CH}}-CH_2-CH_3$

b. $CH_3-\overset{\displaystyle O}{\overset{\|}{C}}-CH_3$ + H_2 $\xrightarrow{Ni}$ $CH_3-\overset{\displaystyle OH}{\overset{|}{CH}}-CH_3$

c. $CH_3-\overset{\displaystyle O}{\overset{\|}{C}}-CH_3$ + 1 CH_3-OH $\rightleftharpoons$ $CH_3-\overset{\displaystyle OH}{\underset{\displaystyle O-CH_3}{\overset{|}{\underset{|}{C}}}}-CH_3$

d. $CH_3-CH_2-\overset{\displaystyle O}{\overset{\|}{C}}-CH_3$ $\xrightarrow{\text{mild oxidizing agent}}$ no reaction

e. $CH_3-CH_2-\overset{\displaystyle O}{\overset{\|}{C}}-CH_3$ + 2 CH_3-CH_2-OH $\rightleftharpoons$ $CH_3-CH_2-\overset{\displaystyle O-CH_2-CH_3}{\underset{\displaystyle O-CH_2-CH_3}{\overset{|}{\underset{|}{C}}}}-CH_3$

13.9

Compounds	Higher boiling point	Explain your choice
pentanal and hexane	pentanal	Pentanal has a higher boiling point because of dipole-dipole attractions.
2-hexanone and octane	2-hexanone	2-Hexanone has a higher boiling point because of dipole-dipole attractions.
pentanal and 1-pentanol	1-pentanol	1-Pentanol has hydrogen bonding between molecules.

13.10 a.   2-methylpropanoic acid
b.   4-iodobenzoic acid
c.   4-chloro-3-methylpentanoic acid

13.11

a. 3-chloro-2-methylpropanoic acid	b. 2,2-dibromobutanoic acid
$CH_2-\overset{\displaystyle }{\underset{\displaystyle Cl}{\overset{|}{C}}H}-\overset{\displaystyle }{\underset{\displaystyle CH_3}{\overset{|}{}}}\overset{\displaystyle O}{\overset{\|}{C}}-OH$	$CH_3-CH_2-\overset{\displaystyle Br}{\underset{\displaystyle Br}{\overset{|}{\underset{|}{C}}}}-\overset{\displaystyle O}{\overset{\|}{C}}-OH$

13.12

IUPAC name	Common name	Structural formula
ethanoic acid	acetic acid	$CH_3-\overset{\overset{\displaystyle O}{\|\|}}{C}-OH$
methanoic acid	formic acid	$H-\overset{\overset{\displaystyle O}{\|\|}}{C}-OH$
butanedioic	succinic acid	$HO-\overset{\overset{\displaystyle O}{\|\|}}{C}-CH_2-CH_2-\overset{\overset{\displaystyle O}{\|\|}}{C}-OH$

13.13

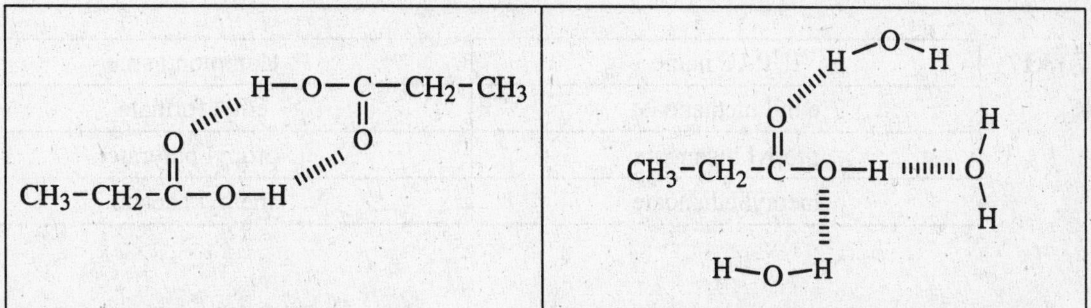

a. two molecules of propanoic acid

b. one molecule of propanoic acid and water

13.14

Name of acid	Structural formula of carboxylic acid salt	IUPAC name of carboxylic acid salt
propanoic acid	$CH_3-CH_2-\overset{\overset{\displaystyle O}{\|\|}}{C}-O^- \; Na^+$	sodium propionate
butanoic acid	$CH_3-CH_2-CH_2-\overset{\overset{\displaystyle O}{\|\|}}{C}-O^- \; Na^+$	sodium butanoate
benzoic acid	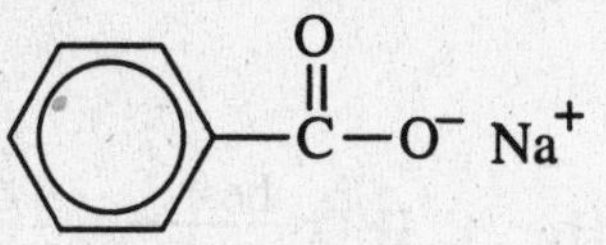	sodium benzoate

13.15 $CH_3-CH_2-\overset{\overset{\displaystyle O}{\|\|}}{C}-O^- \; Na^+ \; + \; HCl \; \longrightarrow \; CH_3-CH_2-\overset{\overset{\displaystyle O}{\|\|}}{C}-OH \; + \; NaCl$

13.16

Acid and alcohol	Structure of ester formed	IUPAC name of ester
ethanoic acid and methanol	$$CH_3-\overset{\overset{\displaystyle O}{\|}}{C}-O-CH_3$$	methyl ethanoate
butanoic acid and methanol	$$CH_3-CH_2-CH_2-\overset{\overset{\displaystyle O}{\|}}{C}-O-CH_3$$	methyl butanoate
benzoic acid and ethanol	$$\bigcirc\!\!\!\!\!\!-\overset{\overset{\displaystyle O}{\|}}{C}-O-CH_2-CH_3$$	ethyl benzoate

13.17

IUPAC name	Common name
ethyl methanoate	ethyl formate
propyl butanoate	propyl butyrate
methyl ethanoate	methyl acetate

13.18 $\quad CH_3-CH_2-CH_2-\overset{\overset{\displaystyle O}{\|}}{C}-O-CH_2-CH_3 \overset{H^+}{\rightleftharpoons} CH_3-CH_2-CH_2-\overset{\overset{\displaystyle O}{\|}}{C}-OH + CH_3-CH_2-OH$

13.19 a. $\quad CH_3-CH_2-CH_2-\overset{\overset{\displaystyle O}{\|}}{C}-NH_2 + H_2O \overset{heat}{\longrightarrow} CH_3-CH_2-CH_2-\overset{\overset{\displaystyle O}{\|}}{C}-OH + NH_3$

b. $\quad \underset{\underset{\displaystyle Cl}{\|}}{CH_3-CH}-\overset{\overset{\displaystyle O}{\|}}{C}-\underset{\underset{\displaystyle CH_3}{}}{N}-CH_2-CH_3 + H_2O \overset{heat}{\longrightarrow} \underset{\underset{\displaystyle Cl}{\|}}{CH_3-CH}-\overset{\overset{\displaystyle O}{\|}}{C}-OH + CH_3-CH_2-\overset{\overset{\displaystyle CH_3}{\|}}{N}H$

c. $\quad \underset{\underset{\displaystyle Br}{}}{\bigcirc}-\overset{\overset{\displaystyle O}{\|}}{C}-\overset{\overset{\displaystyle H}{\|}}{N}-CH_3 + H_2O \overset{heat}{\longrightarrow} \underset{\underset{\displaystyle Br}{}}{\bigcirc}-\overset{\overset{\displaystyle O}{\|}}{C}-OH + CH_3-NH_2$

13.20

Structural formula	Primary, secondary, or tertiary amide	IUPAC name of amide
O‖ CH₃–C–NH–CH₃	secondary	*N*-methylethanamide
O‖ CH₃–CH₂–CH₂–C–NH₂	primary	butanamide
(benzene ring)–C(=O)–N(CH₃)–CH₃	tertiary	*N,N*-dimethylbenzamide

13.21

CH₃–CH₂–CH₂–CH₂–C(=O)–N(CH₂–CH₂–CH₃)–CH₂–CH₃	(benzene ring)–C(=O)–N(CH₃)–CH₃
a. *N*-ethyl-*N*-propylpentamide	b. *N,N*-dimethylbenzamide

13.22 a.   CH₃–CH₂–CH₂–C(=O)–NH(CH₃) + H–OH $\xrightarrow{heat}$ CH₃–CH₂–CH₂–C(=O)–OH + NH₂(CH₃)

b.   (benzene ring)–C(=O)–N(CH₂–CH₃)–CH₃ + H–OH $\xrightarrow{heat}$ (benzene ring)–C(=O)–OH + HN(CH₂–CH₃)–CH₃

c.   CH₃–CH₂–C(=O)–NH₂ + H–OH $\xrightarrow{heat}$ CH₃–CH₂–C(=O)–OH + NH₃

13.23   a. acetal C.
b. aldehyde I.
c. amide D.
d. carboxylate salt E.
e. carboxylic acid F.
f. ester H.
g. ketal A.
h. hemiacetal G.
i. ketone B.

Answers to Self-Test

The numbers in parentheses refer to sections in your textbook:
1. F; three (13.4) **2.** T (13.2) **3.** T (13.5) **4.** F; lower (13.5) **5.** T (13.3) **6.** F; reduced (13.5)
7. T (13.3) **8.** F; formic acid (13.6) **9.** F; hexanoic acid (13.6) **10.** T (13.7)
11. F; high boiling points (13.7) **12.** T (13.7) **13.** F; esters (13.9) **14.** T (13.9)
15. F; a condensation polymer (13.12) **16.** T (13.12) **17.** T (13.12) **18.** F; one (13.10) **19.** d (13.2)
20. a (13.4) **21.** c (13.5) **22.** a (13.4) **23.** a (13.8) **24.** c (13.8) **25.** c (13.9) **26.** a (13.6) **27.** b (13.9)
28. c (13.11)

Chapter Overview

The remaining chapters in the book will be concerned with compounds that are important in biological systems. Carbohydrates are important energy-storage compounds. Their oxidation provides energy for the activities of living organisms.

In this chapter you will learn to identify various types of carbohydrates and write structural diagrams for some of the most common ones. You will be able to explain the concept of handedness in terms of chiral carbons, and draw figures that represent the three-dimensional structures of compounds containing chiral carbons. You will identify the structural features of some important disaccharides and polysaccharides, and learn where in nature they are found and what functions they have.

Practice Exercises

14.1 **Monosaccharides** (Sec. 14.3) are **carbohydrates** (Sec. 14.3) that contain a single polyhydroxy aldehyde or polyhydroxy ketone unit. Carbohydrates can be classified by the number of monosaccharide units they contain. Give the names for carbohydrates having the following numbers of monosaccharide units.

   a. 2 monosaccharide units   _____

   b. 2 to 10 monosaccharide units _____

   c. 3 monosaccharide units   _____

   d. many (greater than 10) units _____

14.2 An object that cannot be superimposed on its **mirror image** (Sec. 14.5) is a **chiral object** (Sec. 14.5). An organic molecule is chiral if it contains a **chiral center** (Sec. 14.5), a single atom with four different atoms or groups of atoms attached to it. Indicate whether the circled carbon in each structure below is chiral or **achiral** (Sec. 14.5):

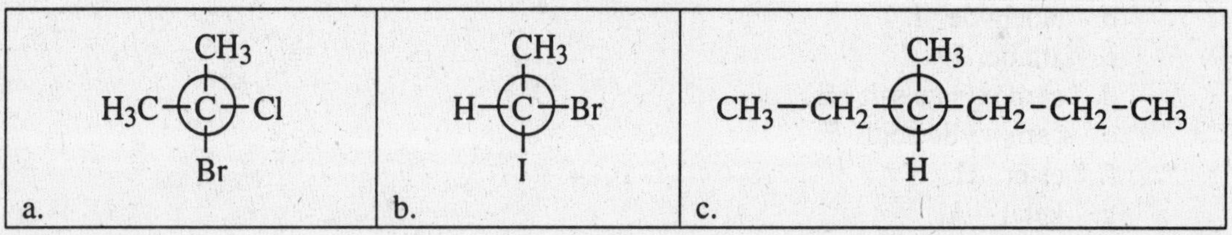

CH_3	CH_3	CH_3
H_3C—Ⓒ—Cl	H—Ⓒ—Br	CH_3—CH_2—Ⓒ—CH_2-CH_2-CH_3
Br	I	H
a.	b.	c.

14.3 Organic molecules may contain more than one chiral center. Circle the chiral centers in the following condensed structures.

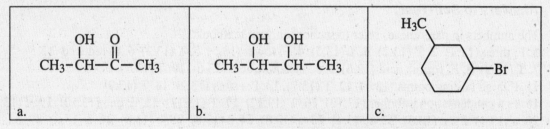

OH O	OH OH	H_3C
CH_3–CH—C—CH_3	CH_3–CH—CH—CH_3	⬡—Br
a.	b.	c.

14.4 The three-dimensional nature of chiral carbons can be represented by showing the chiral carbon as intersecting horizontal and vertical lines. When the carbonyl group is drawn at or near the top of the structure, the right-handed and left-handed forms, or D- and L-forms, of the molecule depend on the left or right position of –OH on the chiral carbon furthest from the carbonyl group.

Test your ability to interpret this type of diagram by answering these questions for each structure shown below:
1) Is the molecule left-handed or right-handed?
2) Is the molecule in the D-form or the L-form?
3) How many chiral centers are in the molecule?

a.	b.	c.
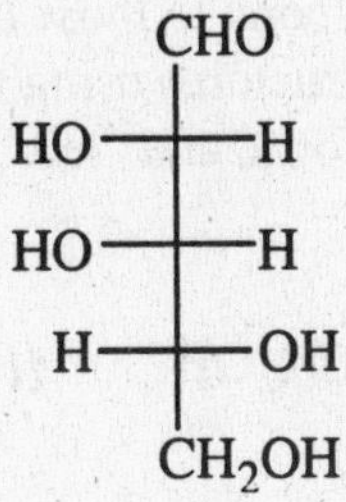 CHO HO——H HO——H H——OH CH₂OH	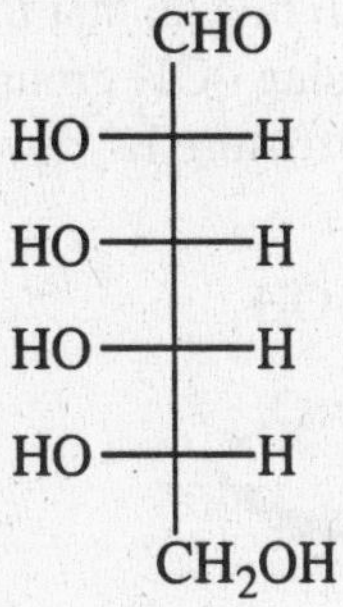 CHO HO——H HO——H HO——H HO——H CH₂OH	CH₂OH C=O H——OH H——OH HO——H CH₂OH
1) 2) 3)	1) 2) 3)	1) 2) 3)

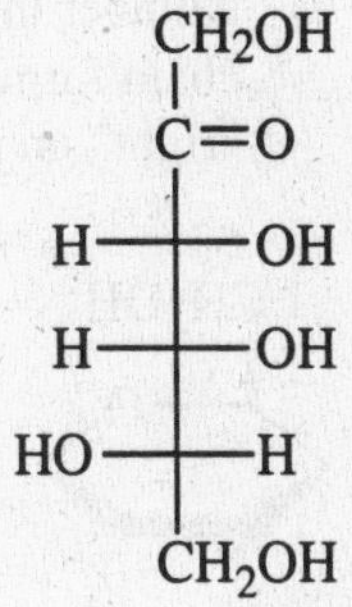

14.5 Monosaccharides are classified as **aldoses** or **ketoses** (Sec. 14.4) on the basis of the carbonyl group: an aldose contains an aldehyde group and a ketose contains a ketone group. They can be further classified by the number of carbons in the molecule (for example, aldopentose). In Sections 14.6 – 14.10 of your textbook you will work with the four representative monosaccharides shown below. All four of these are commonly found in biological systems.

Classify these four monsaccharides by matching the letter of the structure with its description below. Write the name of each monosaccharide below its structural formula and number the carbon atoms in each structure so that the carbonyl carbon has the lowest possible number.

A.	B.	C.	D.
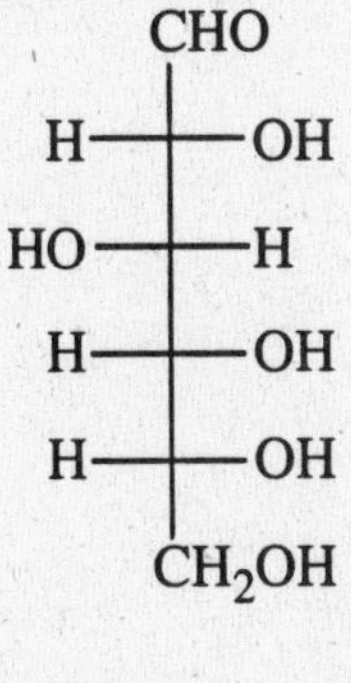 CHO H——OH HO——H H——OH H——OH CH₂OH	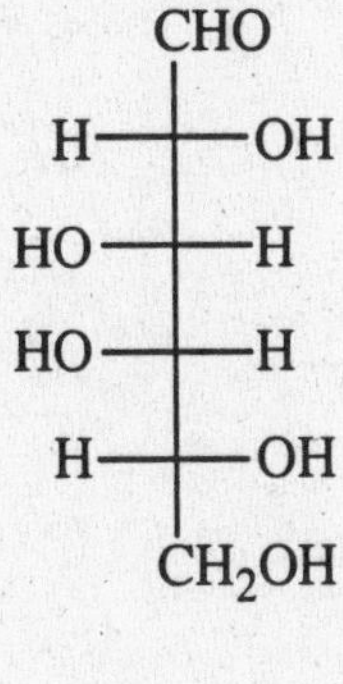 CHO H——OH HO——H HO——H H——OH CH₂OH	CH₂OH C=O HO——H H——OH H——OH CH₂OH	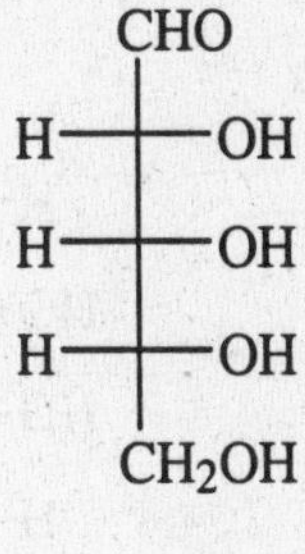 CHO H——OH H——OH H——OH CH₂OH

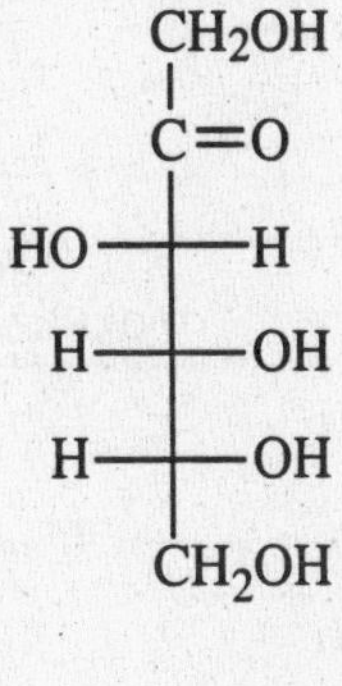

aldohexose _____

aldopentose _____

ketopentose _____

14.6   For monosaccharides containing five or more carbons, open-chain structures are in
equilibrium with cyclic structures. The cyclic structures are formed by the intramolecular
reaction of the carbonyl group with a hydroxyl group (on carbon 5 for a hexose and carbon 4
for a pentose) to form a cyclic hemiacetal or cyclic hemiketal.

The following exercise will help you to become familiar with the cyclic forms of the four
monosaccharides in Practice Exercise 14.5.

1)   First, circle the hemiacetal or hemiketal carbon in each of the cyclic structures below.
(This corresponds to the carbonyl carbon in the open-chain structure.) Write in the
–OH group on this carbon so that the monosaccharide is in the α-form. The rule is: The
–OH is in the "down" position for the α-form and the "up" position for the β-form.

2)   Number the carbons on each ring so that the numbers correspond to those on the open-
chain forms. Fill in the missing –OH groups on the cyclic structures using the rule that
"right" on the open-chain structure is "down" in the cyclic form, and "left" is "up."

α–D-glucose   α–D-galactose   α–D-fructose   α–D-ribose

The α-form and the β-form of a cyclic monsaccharide are in equilibrium with one another.
How would you change the above structural formulas to represent the β-forms of the
monosaccharides? _____

14.7   Weak oxidizing agents oxidize the carbonyl group end of a monsaccharide to give an acid.
Sugars that can be oxidized by these agents are called **reducing sugars** (Sec. 14.8). The
carbonyl group of monosaccharides can be reduced to a hydroxyl group, using H_2 and a
catalyst, to form a polyhydroxy alcohol. Another important reaction of monosaccharides is
glycoside (Sec. 14.8) formation, reaction of the cyclic hemiacetal or hemiketal with an
alcohol to form an acetal or a ketal.

Complete the following equations showing some common reactions of sugars.

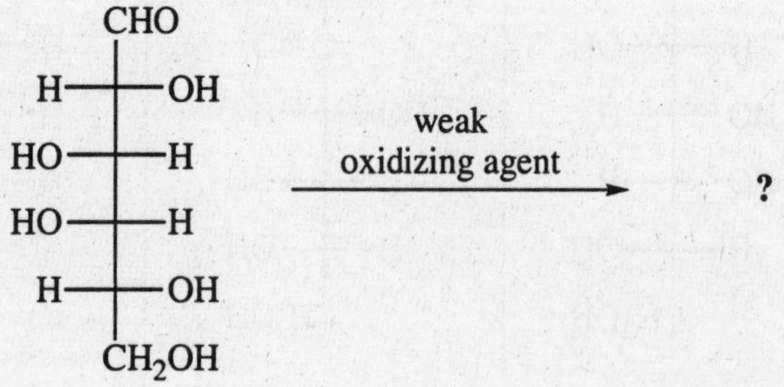

b.

$$\begin{array}{c} \text{CHO} \\ \text{H}\!-\!\text{OH} \\ \text{H}\!-\!\text{OH} \\ \text{H}\!-\!\text{OH} \\ \text{CH}_2\text{OH} \end{array} \xrightarrow[\text{catalyst}]{\text{H}_2} \quad ?$$

c.

$+$ CH_3OH $\xrightleftharpoons{H^+}$ $?$ $+$ $?$

14.8 The two monosaccharides that form a **disaccharide** (Sec. 14.9) are joined by a **glycosidic linkage** (Sec. 14.9). The configuration (α or β) of the hemiacetal carbon atom of the cyclic form is often very important.

For the following disaccharides: a. give the name; b. circle the glycosidic linkage and name the linkage (α or β); c. give the names of the monosaccharides that make up the disaccharide; d. give the common name of the disaccharide.

	a. b. c. d.
	a. b. c. d.
	a. b. c. d.

14.9   There are a number of important **polysaccharides** (Sec. 14.3) that are made up of D-glucose units. They differ in the type of linkage between monomers, the size, and the function.

Complete the following table summarizing these properties. Use information found in Section 14.10 of your textbook.

Polysaccharide	Linkage	Size	Branching	Function
glycogen				
amylose				
amylopectin				
cellulose				

Self-Test

True-false: Indicate whether the following statements are true or false. If the statement is false, give the word or phrase that may be substituted for the underlined portion to make the statement true.

1.   Bioorganic substances include carbohydrates, lipids, proteins, and <u>nucleic acids</u>.

2.   Carbohydrates form part of the structural framework of <u>DNA and RNA</u>.

3.   An oligosaccharide is a carbohydrate that contains <u>at least 20</u> monosaccharide units.

4.   <u>A chiral object</u> is an object that is identical to its mirror image.

5.   The compound 2-methyl-2-butanol contains <u>one</u> chiral center.

6   <u>Proteins</u> are the most abundant bioorganic substances on earth.

7.   A six-carbon monosaccharide with a ketone functional group is called <u>an aldopentose</u>.

8.   Fructose can form a 5-membered ring that is <u>a hemiketal</u>.

9.   The carbonyl group of a monosaccharide can be reduced to a hydroxyl group using <u>Tollens' solution</u> as a reducing agent.

10.   The bond between two monosaccharide units in a disaccharide is a <u>glycosidic</u> linkage.

11.   Humans cannot digest cellulose because they lack an enzyme to catalyze hydrolysis of the <u>$\alpha(1\rightarrow4)$ linkage</u>.

12.   Amylopectin molecules are <u>more highly branched</u> than amylose molecules are.

13.   Chitin is a linear polymer made up of units which are <u>derivatives of glucose.</u>

Multiple choice:

14.   Which of the following substances is not required for the production of carbohydrates by photosynthesis?

    a. carbon dioxide   b. oxygen   c. water
    d. solar energy   e. chlorophyll

15. Which of these compounds contains three chiral centers?

    a. glyceraldehyde   b. glucose   c. fructose
    d. galactose   e. none of these

16. Ribose is an example of a(n):

    a. ketohexose   b. aldohexose   c. aldotetrose
    d. aldopentose   e. none of these

17. Dextrose, or blood sugar, is:

    a. D-glucose   b. D-galactose   c. D-fructose
    d. D-ribose   e. none of these

18. Which of the following compounds is **not** made up solely of D-glucose units?

    a. maltose   b. lactose   c. cellulose
    d. starch   e. glycogen

19. Which of the following sugars is **not** a reducing sugar?

    a. lactose   b. maltose   c. sucrose
    d. galactose   e. all are reducing sugars

20. The main storage form of D-glucose in animal cells is:

    a. glycogen   b. amylopectin   c. amylose
    d. sucrose   e. none of these

21. Which of the following molecules has a chiral center?

    a. ethanol   b. 1-chloro-1-bromoethane   c. 1-chloro-2-bromoethane
    d. 1,2-dichloroethane   e. none of these

22. Hydrolysis of sucrose yields:

    a. glucose and fructose   b. glucose and galactose   c. ribose and fructose
    d. ribose and glucose   e. none of these

Answers to Practice Exercises

14.1   a. disaccharides
         b. oligosaccharides
         c. trisaccharides
         d. polysaccharides

14.2   a. achiral
         b. chiral
         c. chiral

14.3

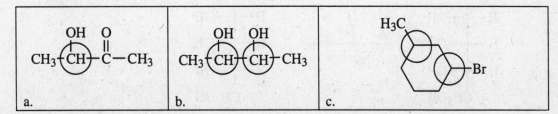

14.4

a.	b.	c.
1) right-handed	1) left-handed	1) left-handed
2) D-form	2) L-form	2) L-form
3) 3 chiral centers	3) 4 chiral centers	3) 3 chiral centers

14.5

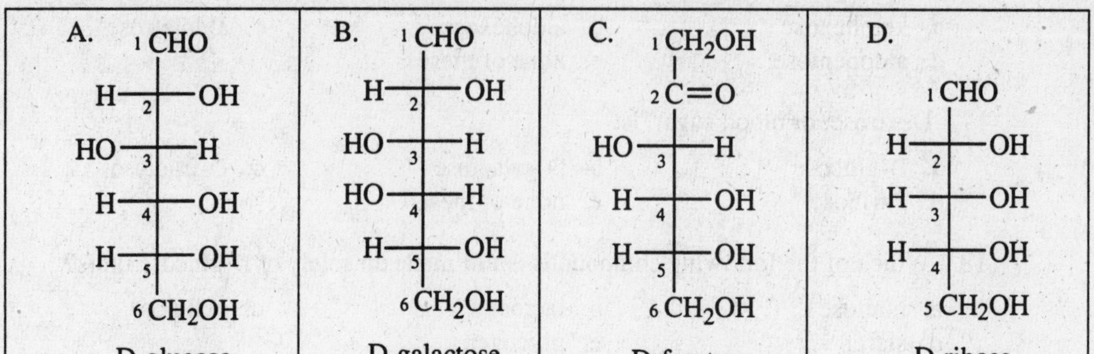

| A. ^{1}CHO
 H —$_2$— OH
 HO —$_3$— H
 H —$_4$— OH
 H —$_5$— OH
 $_6$CH$_2$OH
 D-glucose | B. ^{1}CHO
 H —$_2$— OH
 HO —$_3$— H
 HO —$_4$— H
 H —$_5$— OH
 $_6$CH$_2$OH
 D-galactose | C. $_1$CH$_2$OH
 $_2$C=O
 HO —$_3$— H
 H —$_4$— OH
 H —$_5$— OH
 $_6$CH$_2$OH
 D-fructose | D. $_1$CHO
 H —$_2$— OH
 H —$_3$— OH
 H —$_4$— OH
 $_5$CH$_2$OH
 D-ribose |

aldohexose A and B
aldopentose D
ketopentose C

14.6

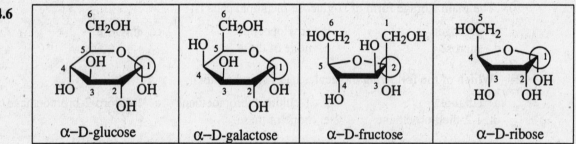

α–D-glucose   α–D-galactose   α–D-fructose   α–D-ribose

How would you change each of these to the β-form? Change the –OH on the carbonyl carbon (circled above) to the "up" position.

14.7

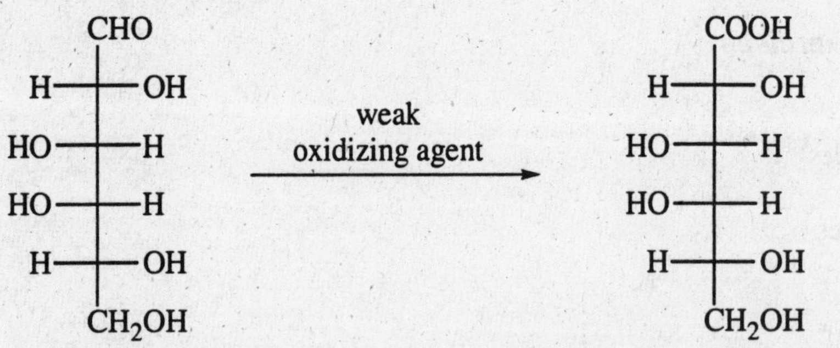

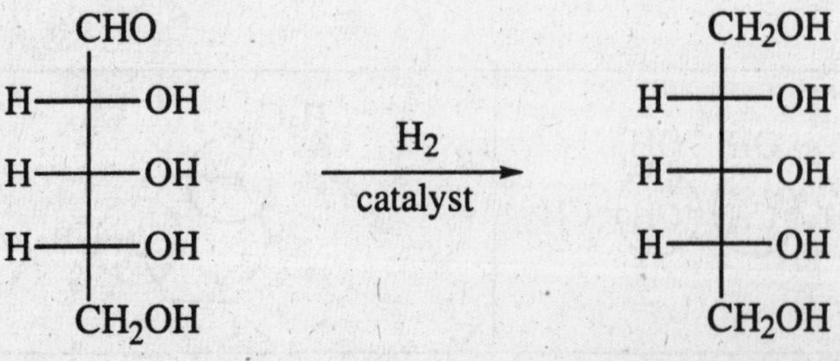

CHO
H —— OH
HO —— H
HO —— H
H —— OH
CH$_2$OH

weak oxidizing agent →

COOH
H —— OH
HO —— H
HO —— H
H —— OH
CH$_2$OH

CHO
H —— OH
H —— OH
H —— OH
CH$_2$OH

H_2 catalyst →

CH$_2$OH
H —— OH
H —— OH
H —— OH
CH$_2$OH

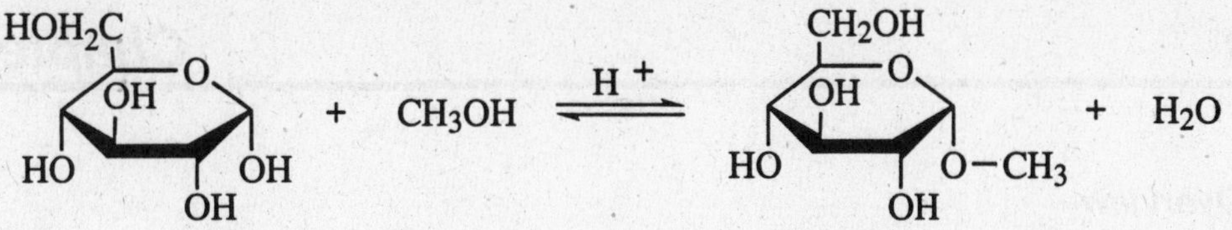

14.8

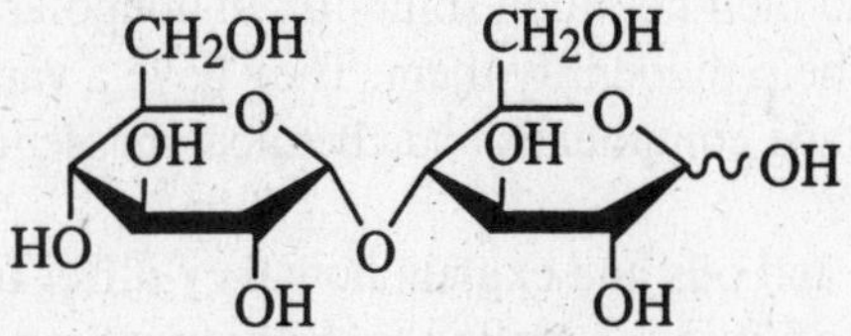

a. maltose
b. α(1 → 4) linkage
c. α-D-glucose and D-glucose
d. malt sugar

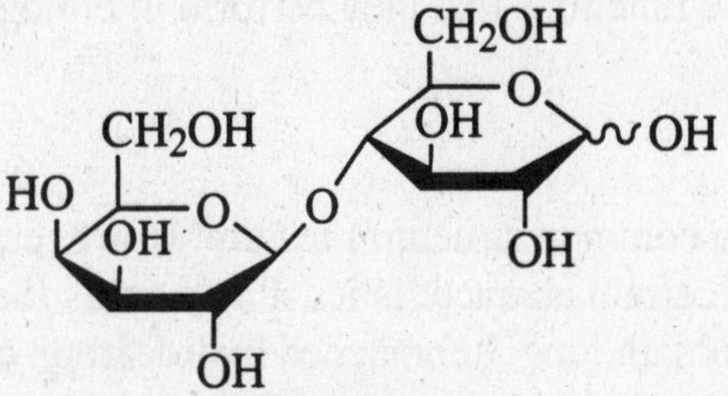

a. lactose
b. β(1 → 4) linkage
c. β-D-galactose and D-glucose
d. milk sugar

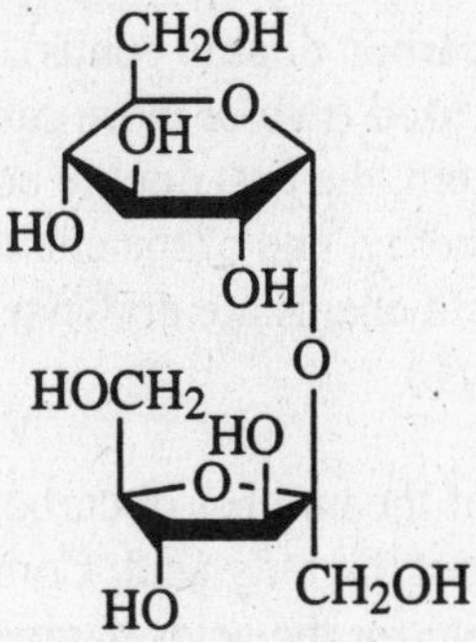

a. sucrose
b. α,β(1 →2)
c. α-D-glucose and β-D-fructose
d. table sugar

14.9

Polysaccharide	Linkage	Size (amu)	Branching	Function
glycogen	α(1→4) α(1→6)	3,000,000	very highly branched	storage form of glucose in animals
amylose	α(1→4)	50,000	straight- chain	storage form of glucose in plants (15-20%)
amylopectin	α(1→4)	300,000	highly branched	storage form of glucose in plants (80-85%)
cellulose	β(1→4)	900,000	linear, unbranched	structural component of cell walls in plants

Answers to Self-Test

The numbers in parentheses refer to sections in your textbook:
1. T (14.1) **2.** T (14.2) **3.** F; two to ten (14.3) **4.** F; an achiral object (14.5) **5.** F; no (14.5)
6. F; carbohydrate (14.2) **7.** F; a ketohexose (14.4) **8.** T (14.7) **9.** F; H_2 and a catalyst (14.8)
10. T (14.9) **11.** F; β(1→4) linkage (14.10) **12.** T (14.10) **13.** T (14.10) **14.** b (14.2) **15.** c (14.5)
16. d (14.4) **17.** a (14.6) **18.** b (14.9) **19.** c (14.9) **20.** a (14.10) **21.** b (14.5) **22.** a (14.9)

Chapter Overview

Lipids are compounds grouped according to their common solubility in nonpolar solvents, but there are some structural features that also help to identify them. They have a variety of functions in the body, acting as energy storage compounds and chemical messengers, and providing structure to cell membranes.

In this chapter you will define fats and oils and explain how they differ in molecular structure. You will identify various groups of lipids according to their structures and components, and you will learn some of the functions that they perform in biological systems.

Practice Exercises

15.1   **Lipids** (Sec. 15.1) as a group do not have a common structural feature. However, there are sub-groups of lipids that that are united by certain characteristics. **Fatty acids** (Sec. 15.2) are naturally occurring carboxylic acids that contain long, unbranched hydrocarbon chains 12 to 26 carbon atoms in length.

Fatty acids are classified as **saturated** (no carbon-carbon double bonds), **monounsaturated** (one carbon-carbon double bond) and **polyunsaturated** (two or more carbon-carbon double bonds) (Sec. 15.2). In the omega classification system, the first double bond is identified by its number on the carbon chain, counted from the methyl end of the chain. (See Table 15.1 in your textbook. Structural formulas for long fatty acid chains are drawn using the notation $(CH_2)_x$, where x is the number of repeated CH_2 units.)

A shorthand notation for fatty acids uses the ratio of the number of carbon atoms to the number of double bonds to express information about the fatty acid. Complete the table below, using the information in the shorthand system for the acids named.

Name of fatty acid	Number of carbon atoms	Number of carbon-carbon double bonds	Omega classification system
myristic acid (14:0)			
linolenic acid (18:3)			
palmitic acid (16:0)			
linoleic acid (18:2)			
oleic acid (18:1)			
stearic acid (18:0)			

15.2   **Triacylglycerols** (Sec. 15.4) are produced by the esterification of three fatty acid molecules with a glycerol molecule.

Draw structural formulas for triacylglycerols with the following fatty acid residues: a. three molecules of oleic acid b. two molecules of stearic acid and one of oleic acid c. three molecules of stearic acid.

a.	b.	c.

15.3   **Fats** (Sec. 15.4) are triacylglycerols with a high percentage of saturated fatty acids; **oils** (Sec. 15.4) are triacylglycerols with a high percentage of unsaturated fatty acids. Tell whether each of the compounds in Practice Exercise 15.2 would be a fat or an oil, and explain.

Compound	Fat or oil?	Explanation
a.		
b.		
c.		

15.4   Triacylglycerols undergo several characteristic reactions. In this exercise you will work with three of these types of reactions and the three triacylglycerol molecules in Practice Exercise 15.2 above.

1.   Hydrolysis is the reverse of esterification. Complete hydrolysis of all three ester linkages in triacylglycerols produces glycerol and three fatty acid molecules. Write the names of the hydrolysis products of the three triacylglycerols in the first column below.

2.   Hydrogenation involves adding H_2 molecules to the double bonds of the fatty acid residues. In the second column, tell how many molecules of hydrogen (H_2) would be used for the complete hydrogenation of each of the three triacylglycerol molecules.

3.   Oxidation of triacylglycerols breaks the carbon-carbon double bonds and produces low molecular weight aldehydes and carboxylic acids that have unpleasant odors. When unsaturated fats have oxidized in this way they are said to be rancid. In the table below, write "yes" in the third column if the triacylglycerol is likely to become rancid, "no" if it is not.

Compound	Hydrolysis products	Moles of hydrogen	Oxidation (rancidity)
a.			
b.			
c.			

15.5   Triacylglycerols are the most common of the lipids that undergo hydrolysis. Other important lipids are the **phosphoacylglycerols** (Sec. 15.6), and the **sphingolipids** (Sec. 15.7). These three types of lipids have similar basic structures, but the component "building blocks" vary.

In the block diagrams below, fill in the name of each of the building blocks of each type of lipid named. Then follow the directions below the diagrams.

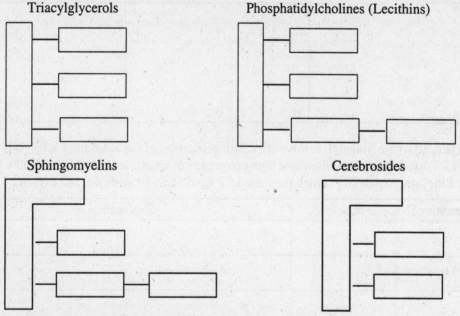

a.   Label all ester linkages in the block diagrams with the letter A.

b.   Label all amide linkages with the letter B.

c.   Label all glycoside linkages with the letter C.

d.   Give the general similarities between triacylglycerols and phosphoacylglycerols

_____

Give the general differences between these two types of lipid. _____

_____

e.   What do all of the lipids in the diagrams above have in common?

_____

15.6   A **steroid** (Sec. 15.8) is a compound based on a characteristic fused-ring system, the steroid nucleus. In the table below match the name of each of the steroids with the number of its function in biological systems.

Answer	Name of lipid	Function of lipid
	androgens	1. pregnancy hormones
	mineralocorticoids	2. control the balance of Na^+ and K^+ ions in cells
	estrogens	3. starting material for synthesis of steroid hormones
	glucocorticoids	4. male sex hormones
	cholesterol	5. emulsifying agents produced by liver
	bile salts	6. female sex hormones
	progestins	7. control glucose metabolism, reduce inflammation

15.7   The aqueous material within a living cell is separated from its surrounding aqueous environment by a **cell membrane** (Sec. 15.9). Most of the mass of the cell membrane is lipid material, in the form of a **lipid bilayer** (Sec. 15.9). Label the diagram of the lipid bilayer shown below, using the letters of the components below the diagram.

Inside the cell

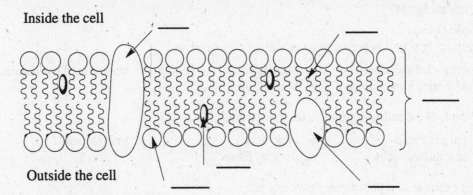

Outside the cell

a.   lipid bilayer
b.   cholesterol molecule
c.   polar heads of membrane molecules
d.   nonpolar tails of membrane molecules
e.   protein molecule that transports nutrients and other substances across the membrane
f.   protein molecule that acts as a receptor for hormones and other substances

Self-Test

True-false: Indicate whether the following statements are true or false. If the statement is false, give the word or phrase that may be substituted for the underlined portion to make the statement true.

1.   Fats contain a high proportion of <u>unsaturated fatty acid</u> chains.

2.   Fats and oils become rancid when the double bonds on triacylglycerol side chains are <u>oxidized</u>.

3.   The two essential fatty acids are <u>linoleic acid and arachidonic acid.</u>

4.   Rancid fats contain <u>low-molecular-mass acids</u>.

5.   <u>Sphingolipids</u> are complex lipids found in the brain and nerves.

6.   In a plasma membrane, the <u>nonpolar tails</u> of the phospholipids are on the outside surfaces of the lipid bilayer.

7.   The presence of cholesterol molecules in the lipid bilayer of a plasma membrane makes the bilayer <u>more flexible</u>.

8.   Essential fatty acids <u>cannot</u> be synthesized within the human body from other substances.

9.   Most naturally occurring triacylglycerols are formed with <u>three identical</u> fatty acid molecules.

10. Partial hydrogenation of vegetable oils produces a product of <u>higher melting point</u> than the original oil.

11. Phosphoacylglycerols are important components of <u>cell membranes</u>.

12. Bile salts play an important role in the human body in the process of <u>digestion</u>.

13. Cholesterol <u>cannot</u> be synthesized within the human body.

14. A fatty acid containing a <u>cis-double bond</u> has a bent carbon chain.

15. Most fatty acids found in the human body have <u>an odd number</u> of carbon atoms.

16. The lipid molecules in the lipid bilayer of a cell membrane are held to one another by <u>covalent bonds</u>.

Multiple choice:

17. A triacylglycerol is prepared by combining glycerol and:

    a. long-chain alcohols   b. fatty acids   c. unsaturated hydrocarbons
    d. saturated hydrocarbons   e. none of these

18. Which of the following fatty acids is saturated?

    a. palmitic acid   b. oleic acid   c. linoleic acid
    d. arachidonic acid   e. none of these

19. An example of a phosphoacylglycerol is:

    a. prostaglandin   b. progesterone   c. a lecithin
    d. glycerol tristearate   e. none of these

20. An example of a steroidal hormone is:

    a. progesterone   b. cholesterol   c. stearic acid
    d. a cerebroside   e. none of these

21. Cholesterol is the starting material within the human body for the synthesis of

    a. Vitamin A   b. Vitamin B1   c. Vitamin C
    d. Vitamin D   e. Vitamin E

22. The most abundant steroid in the human body is:

    a. estradiol   b. testosterone   c. progesterone
    d. cholesterol   e. estrogen

Answers to Practice Exercises

15.1

Name of fatty acid	Number of carbon atoms	Number of carbon-carbon double bonds	Omega classification system
myristic acid (14:0)	14	0	(saturated)
linolenic acid (18:3)	18	3	omega-3
palmitic acid (16:0)	16	0	(saturated)
linoleic acid (18:2)	18	2	omega-6
oleic acid (18:1)	18	1	omega-9
stearic acid (18:0)	18	0	(saturated)

15.2 a.

$$CH_2-O-\overset{\overset{\displaystyle O}{\|}}{C}\!\!\left(CH_2\right)_{\!7}\!\!-\overset{\overset{\displaystyle H}{|}}{C}=\overset{\overset{\displaystyle H}{|}}{C}\!\!\left(CH_2\right)_{\!7}\!\!CH_3$$

$$CH-O-\overset{\overset{\displaystyle O}{\|}}{C}\!\!\left(CH_2\right)_{\!7}\!\!-\overset{\overset{\displaystyle H}{|}}{C}=\overset{\overset{\displaystyle H}{|}}{C}\!\!\left(CH_2\right)_{\!7}\!\!CH_3$$

$$CH_2-O-\overset{\overset{\displaystyle O}{\|}}{C}\!\!\left(CH_2\right)_{\!7}\!\!-\overset{\overset{\displaystyle H}{|}}{C}=\overset{\overset{\displaystyle H}{|}}{C}\!\!\left(CH_2\right)_{\!7}\!\!CH_3$$

b.

$$CH_2-O-\overset{\overset{\displaystyle O}{\|}}{C}\!\!\left(CH_2\right)_{\!16}\!\!CH_3$$

$$CH-O-\overset{\overset{\displaystyle O}{\|}}{C}\!\!\left(CH_2\right)_{\!16}\!\!CH_3$$

$$CH_2-O-\overset{\overset{\displaystyle O}{\|}}{C}\!\!\left(CH_2\right)_{\!7}\!\!-\overset{\overset{\displaystyle H}{|}}{C}=\overset{\overset{\displaystyle H}{|}}{C}\!\!\left(CH_2\right)_{\!7}\!\!CH_3$$

c.

$$CH_2-O-\overset{\overset{\displaystyle O}{\|}}{C}\!\!\left(CH_2\right)_{\!16}\!\!CH_3$$

$$CH-O-\overset{\overset{\displaystyle O}{\|}}{C}\!\!\left(CH_2\right)_{\!16}\!\!CH_3$$

$$CH_2-O-\overset{\overset{\displaystyle O}{\|}}{C}\!\!\left(CH_2\right)_{\!16}\!\!CH_3$$

15.3

Compound	Fat or oil?	Explanation
a.	oil	All three fatty acid residues are unsaturated, so that the molecules are bent and do not pack together closely, resulting in low melting point.
b.	fat	Only one fatty acid is unsaturated and two are saturated, so the molecules are less distorted than in the previous example, and pack together more closely.
c.	fat	All three fatty acids are saturated and linear, resulting in close packing and high melting point.

15.4

Compound	Hydrolysis products	Moles of hydrogen	Oxidation (rancidity)
a.	glycerol, 3 molecules of oleic acid	3	yes
b.	glycerol, 1 molecule of oleic acid, 2 molecules of stearic acid	1	yes
c.	glycerol, 3 molecules of stearic acid	0	no

15.5   Triacylglycerols   Phosphatidylcholines (Lecithins)

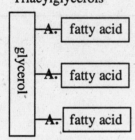   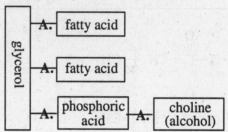

Sphingomyelins   Cerebrosides

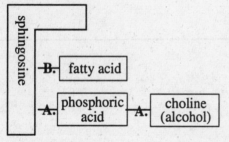   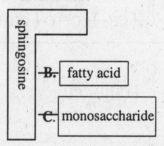

A. ≡ ester linkage   **B.** ≡ amide linkage   **C.** ≡ glycoside linkage

d.   Similarities: polar heads, nonpolar tails (fatty acid chains)

Differences: phosphoacylglycerols contain phosphoric acid and alcohol (in addition to the glycerol and fatty acids.)

e.   They all have in common: polar heads, nonpolar tails, fatty acids, ester linkages.

15.6

Answer	Name of lipid	Function of lipid
4.	androgens	1. pregnancy hormones
2.	mineralocorticoids	2. control the balance of Na^+ and K^+ ions in cells
6.	estrogens	3. starting material for synthesis of steroid hormones
7.	glucocorticoids	4. male sex hormones
3.	cholesterol	5. emulsifying agents produced by liver
5.	bile salts	6. female six hormones
1.	progestins	7. control glucose metabolism, reduce inflammation

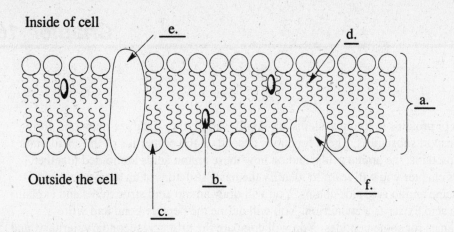

Inside of cell

e.

d.

a.

Outside the cell

b.

f.

c.

Answers to Self-Test

The numbers in parentheses refer to sections in your textbook:

1. F; saturated fatty acid (15.4) **2.** T (15.4) **3.** F; linoleic and linolenic (15.2)

4. T (15.5) **5.** T (15.7) **6.** F; polar heads (15.9) **7.** F; more rigid (15.9) **8.** T (15.2)

9. F; a mixture of (15.4) **10.** T (15.5) **11.** T (15.9) **12.** T (15.8) **13.** F; can (15.8)

14. T (15.2) **15.** F; an even number (15.2) **16.** F; dipole-dipole interactions (15.9) **17.** b (15.4)

18. a (15.2) **19.** c (15.7) **20.** a (15.8) **21.** d (15.8) **22.** d (15.8)

Chapter Overview

The functions of proteins in living systems are highly varied; they catalyze reactions, form structures, transport substances. These functions depend on the properties of the small units that make up proteins, the amino acids, and on how these amino acids are joined together.

In this chapter you will learn to identify the basic structure of an amino acid and characterize some amino acid side chains. You will draw amino acid strucxtures and explain why an amino acid exists as a zwitterion. You will define the peptide bond and write abbreviated names for some peptides. You will compare the primary, secondary, tertiary, and quaternary structures of proteins.

Practice Exercises

16.1 **Proteins** (Sec. 16.1) are polymers in which the monomer untis are amino acids. An α-**amino acid** (Sec. 16.2) contains an amino group and a carboxyl group, both attached to the same carbon atom, called the alpha carbon atom. Amino acids are classified as **nonpolar, polar neutral, polar basic, or polar acidic** (Sec. 16.2), depending on the nature of the side chain.

Complete the following table for classification of some **standard amino acids** (Sec. 16.2).

Name	Abbreviation	Structure	Classification
leucine			
aspartic acid			
serine			
histidine			
asparagine			

Essential amino acids (Sec. 16.2) are those amino acids necessary for the synthesis of human protein, but that cannot be produced in the human body. They must be obtained from food. In the table above, put a star by the name of each essential amino acid.

16.2   The α-carbon atom of an α-amino acid is a chiral center (except in glycine); naturally
occurring amino acids (with a few exceptions) are L isomers.

Draw structural formulas for the following α-amino acids. The "handedness" of an amino
acid can be shown in the same way as for the monosaccharides: draw the –COOH group at
the top of the structure and the –NH$_2$ group to the left of the chiral carbon for the L isomer.
See Table 16.1 in your textbook for side chains.

L-alanine	L-cysteine	L-aspartic acid

16.3   Since an amino acid has both an acidic group (the carboxyl group) and a basic group (the
amino group), it exists as a **zwitterion** (Sec. 16.4), a molecule that has a positive charge on
one atom and a negative charge on another atom. Draw the structural formula for neutral
phenylalanine and the zwitterionic structure of phenylalanine.

Structural formula of phenylalanine (un-ionized form)	Zwitterionic structure of phenylalanine

16.4   In solution, three different amino acid forms exist in equilibrium. The amount of each form
present depends on the pH of the solution.

Draw the structure of the predominant form of phenylalanine at each of the following pHs.

pH = 1	pH = 7	pH = 12

16.5 . The **peptide bond** (Sec. 16.5) is an amide linkage joining amino acids to form **peptides** (Sec. 16.5). The peptide bond forms between the carboxyl group of one amino acid and the amino group of another with the removal of a water molecule.

    a.   Draw the structural formulas of the two different dipeptides that could form from the amino acids alanine (Ala) and isoleucine (Iso).

    b.   Circle the peptide bond in each structure.

    c.   Label the *N*-terminal end and the *C*-terminal end of each peptide chain

16.6 Peptides that contain the same amino acids but in different order are different molecules with different properties.

    a.   Using the three-letter abbreviations, draw the abbreviated peptide formulas for all possible tripeptides that could be formed from two molecules of Pro and one molecule of Lys.

    b.   Using the three-letter abbreviations for the amino acids, draw all possible tripeptides that could be formed from one molecule each of the following amino acids: Pro, Trp, and Lys.

16.7 A polypeptide contains many amino acids linked by peptide bonds. A **protein** (Sec. 16.6) is a polypeptide that contains more than 50 amino acid residues. The sequence of the amino acids in a protein's peptide chain is called its **primary structure** (Sec. 16.7).

The shape of a protein molecule is the result of forces acting between various parts of the peptide chain. These forces determine the **secondary, tertiary, and quaternary structure** of the protein (Sec. 16.7 –16.10).

Complete the following table by writing the type of protein structure (primary, secondary, tertiary, or quaternary) most readily associated with the following terms, and the type of bond interaction that maintains that structure level.

Structural term	Type of protein structure	Type of bonds or interactions
amino acid sequence		
β-pleated sheet		
disulfide bonds		
α-helix		
globular protein (Sec. 16.11)		
hemoglobin (tetramer)		
fibrous protein (Sec. 16.11)		
protein hydrolysis		
denaturation		

16.8   **Enzymes** (Sec. 16.13) are catalysts for biological reactions. They are commonly named by taking the name of the **substrate** (Sec. 16.14), the compound undergoing change, and adding the ending -*ase*. Sometimes the enzyme name gives the type of reaction being catalyzed.

In the table below, predict the function for a given enzyme name by matching the name with the function and/or substrate

Answer	Name of enzyme	Type of reaction
	α-amylase	1. removal of a carboxyl group from a substrate
	dehydrogenase	2. oxidation of a substrate
	decarboxylase	3. hydrolysis of ester linkages in lipids
	transaminase	4. hydrolysis of peptide linkages
	peptidase	5. hydrolysis of α-linkages in starch molecules
	oxidase	6. transfer of an amino group from one molecule to another
	lipase	7. removal of hydrogen from a substrate

16.9   Enzymes can be divided into two classes: **simple enzymes** and **conjugated enzymes** (Sec. 16.14). Other common terms used in describing enzymes are defined in Section 16.14 of your textbook. Match the terms below with the best description.

Answer	Term	Description
	coenzyme	1. nonprotein portion of a conjugated enzyme
	cofactor	2. another name for conjugated enzyme
	simple enzyme	3. protein portion of a conjugated enzyme
	apoenzyme	4. small organic molecule that serves as a cofactor
	holoenzyme	5. composed only of protein

16.10   Before an enzyme-catalyzed reaction takes place, an **enzyme-substrate complex** (Sec. 16.15) is formed: the substrate binds to the **active site** (Sec. 16.15) of the enzyme.

a.   What are two models that account for the specific way an enzyme selects a substrate?

b.   What is the main difference between these two models?

16.11   **Enzyme activity** (Sec. 16.16) is a measure of the rate at which an enzyme converts substrate to products.

a.   Name four factors that affect enzyme activity.

b.   Describe the effects on rate of reaction of an increase in each of the four factors named above.

Self-Test

True-false: Indicate whether the following statements are true or false. If the statement is false, give the word or phrase that may be substituted for the underlined portion to make the statement true.

1.   Naturally occurring amino acids are generally in the L form.

2.   A peptide bond is the bond formed between the amino group of an amino acid and the carboxyl group of the same amino acid.

3.   Since amino acids have both an acidic group and a basic group, they are able to undergo internal acid-base reactions.

4.   When the pH of an amino acid solution is lowered, the amino acid zwitterion forms more of the positively charged species.

5.   A peptide bond differs slightly from an amide bond.

6.   A tripeptide has a COO^- group at each end of the molecule.

7.   Next to water, proteins are the most abundant substances in cells.

8.   The function of a protein is controlled by the protein's primary structure.

9.   The alpha helix structure is a part of the primary structure of some proteins.

10.   Denaturation of a protein involves changes in the protein's primary structure.

11.   The quaternary structure of a protein involves associations among separate polypeptide chains.

12.   The peptides in a beta-pleated sheet are held in place by hydrogen bonds.

13. In an enzyme catalyzed reaction, the compound that undergoes a chemical change is called a <u>substrate</u>.

14. Enzyme names are usually based on the <u>structure</u> of the enzyme.

15. The protein portion of a conjugated enzyme is called the <u>coenzyme</u>.

16. The <u>active site</u> of an enzyme is the small part of an enzyme where catalysis takes place.

17. According to the lock-and-key model of enzyme action, the active site of the enzyme is <u>flexible</u> in shape.

18. A carboxypeptidase is an enzyme that is specific for <u>one group of compounds</u>.

19. A cofactor is a <u>protein part</u> of an enzyme necessary for the enzyme's function.

Multiple choice:

20. An example of a polar basic amino acid is:

    a. lysine   b. serine   c. tryptophan
    d. leucine   e. none of these

21. A tripeptide is formed from two alanine molecules and one glycine molecule. The maximum number of different tripeptides that could be formed from this combination is:

    a. two   b. three   c. four
    d. five   e. none of these

22. Which of the following attractive interactions does **not** affect the formation of a protein's tertiary structure?

    a. disulfide bonds   b. salt bridges   c. hydrogen bonds
    d. peptide bonds   e. none of these

23. Proteins may be denatured by:

    a. heat   b. acid
    c. ethanol   d. all of the above (a, b, and c)
    e. a and b only

24. Which of these is true of globular proteins?

    a. stringlike molecules   b. water-insoluble
    c. shape is roughly spherical d. structural function in body
    e. none of these

25. Enzymes assist chemical reactions by:

    a. increasing the rate of the reactions
    b. increasing the temperature of the reactions
    c. being consumed during the reactions
    d. all of these
    e. none of these

26. The enzyme that catalyzes the reaction of an alcohol to form an aldehyde would be:

    a. an oxidase   b. a decarboxylase   c. a dehydratase
    d. a reductase   e. none of these

Answers to Practice Exercises

16.1

Name	Abbreviation	Structure	Classification
leucine	Leu	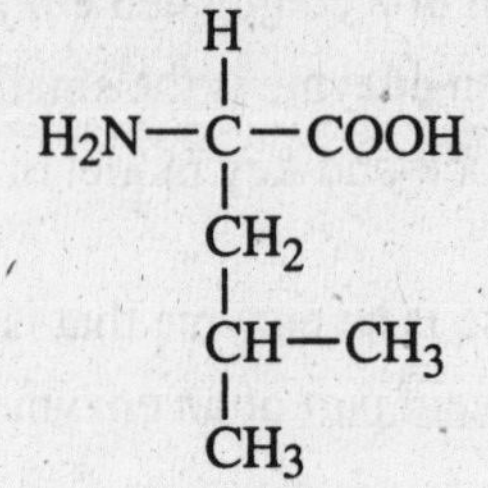 H₂N—C—COOH with H above, CH₂, CH—CH₃, CH₃	nonpolar
aspartic acid	Asp	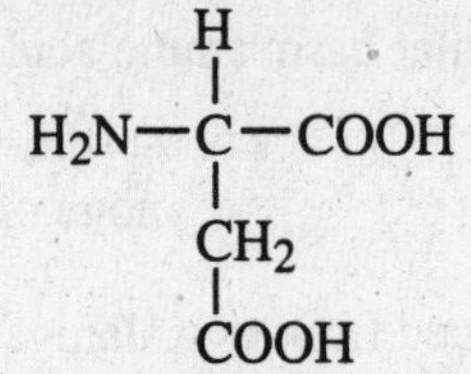 H₂N—C—COOH with H above, CH₂, COOH	polar acidic
serine	Ser	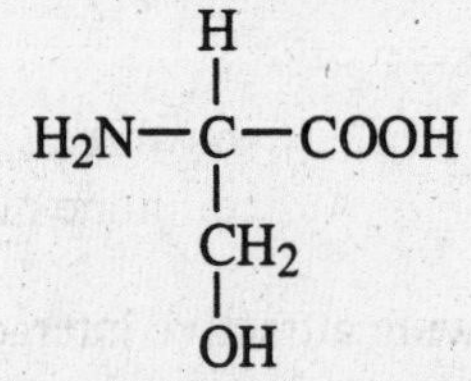 H₂N—C—COOH with H above, CH₂, OH	polar neutral
histidine	His	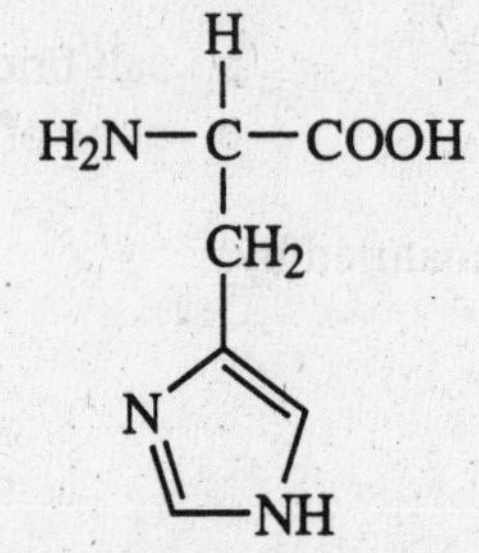 H₂N—C—COOH with H above, CH₂, imidazole ring (N, NH)	polar basic
asparagine	Asn	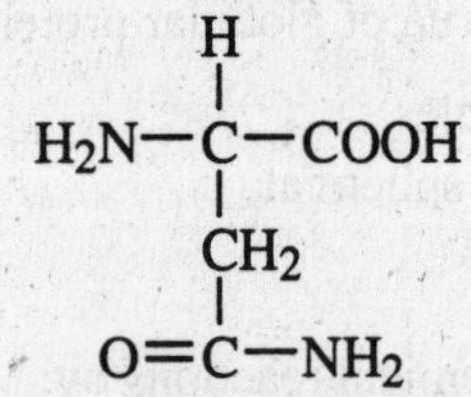 H₂N—C—COOH with H above, CH₂, O=C—NH₂	polar neutral

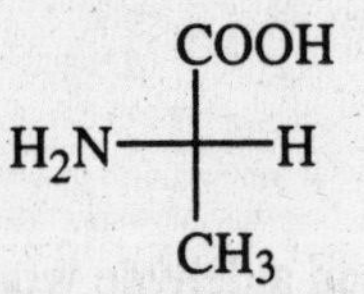

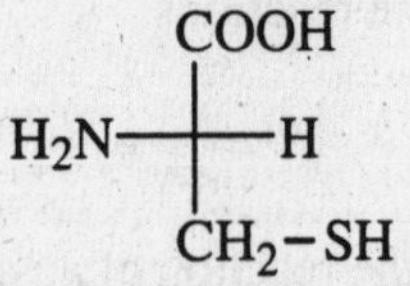

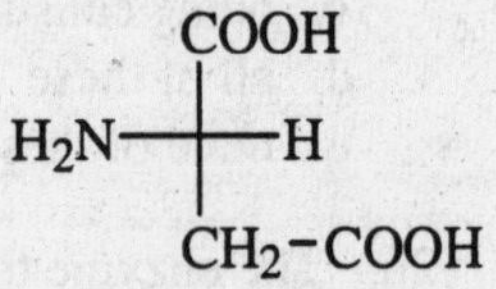

16.2

COOH, H₂N—C—H, CH₃ **L-alanine**	COOH, H₂N—C—H, CH₂—SH **L-cysteine**	COOH, H₂N—C—H, CH₂—COOH **L-aspartic acid**

16.3

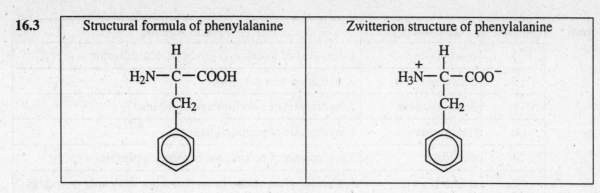

Structural formula of phenylalanine	Zwitterion structure of phenylalanine
$H_2N-\overset{\overset{\displaystyle H}{\displaystyle \vert}}{\underset{\underset{\displaystyle \text{(phenyl ring)}}{\displaystyle \vert}}{\underset{\displaystyle CH_2}{C}}}-COOH$	$\overset{+}{H_3N}-\overset{\overset{\displaystyle H}{\displaystyle \vert}}{\underset{\underset{\displaystyle \text{(phenyl ring)}}{\displaystyle \vert}}{\underset{\displaystyle CH_2}{C}}}-COO^-$

16.4

pH = 1: $\overset{+}{H_3N}-\overset{H}{\underset{CH_2}{C}}-COOH$ (phenyl ring)

pH = 7: $\overset{+}{H_3N}-\overset{H}{\underset{CH_2}{C}}-COO^-$ (phenyl ring)

pH = 12: $H_2N-\overset{H}{\underset{CH_2}{C}}-COO^-$ (phenyl ring)

16.5

N-terminal end ... $\overset{+}{H_3N}-\overset{H}{\underset{CH_3}{C}}-\overset{O}{C}-\overset{H}{N}-\overset{H}{\underset{\underset{CH_3}{\underset{CH_2}{CH-CH_3}}}{C}}-COO^-$ ... C-terminal end

N-terminal end ... $\overset{+}{H_3N}-\overset{H}{\underset{\underset{CH_3}{\underset{CH_2}{CH-CH_3}}}{C}}-\overset{O}{C}-\overset{H}{N}-\overset{H}{\underset{CH_3}{C}}-COO^-$ ... C-terminal end

16.6   a. Pro-Pro-Lys, Pro-Lys-Pro, Lys-Pro-Pro

b. Pro-Trp-Lys, Pro-Lys-Trp, Trp-Pro-Lys, Trp-Lys-Pro, Lys-Trp-Pro, Lys-Pro-Trp.

16.7

Structural term	Type of protein structure	Type of bonds or interactions
amino acid sequence	primary	peptide bonds
β-pleated sheet	secondary	hydrogen bonds
disulfide bonds	tertiary	covalent bonds
α-helix	secondary	hydrogen bonds
globular protein (Sec. 16.11)	tertiary	hydrophobic interactions
hemoglobin (tetramer)	quaternary	hydrophobic interactions
fibrous protein (Sec. 16.11)	secondary	hydrogen bonds
protein hydrolysis	primary	peptide bonds
denaturation	secondary, tertiary, and quaternary	side chain interactions (covalent bonds, hydrogen bonding etc.)

16.8

Answer	Name of enzyme	Type of reaction
5.	α-amylase	1. removal of a carboxyl group from a substrate
7.	dehydrogenase	2. oxidation of a substrate
1.	decarboxylase	3. hydrolysis of ester linkages in lipids
6.	transaminase	4. hydrolysis of peptide linkages
4.	peptidase	5. hydrolysis of α-linkages in starch molecules
2.	oxidase	6. transfer of an amino group from one molecule to another
3.	lipase	7. removal of hydrogen from a substrate

16.9

Answer	Term	Description
4.	coenzyme	1. nonprotein portion of a conjugated enzyme
1.	cofactor	2. another name for conjugated enzyme
5.	simple enzyme	3. protein portion of a conjugated enzyme
3.	apoenzyme	4. small organic molecule that serves as a cofactor
2.	holoenzyme	5. composed only of protein

16.10 a. The two models are the lock-and-key model and the induced-fit model.

b. The main difference between the two models is that according to the lock-and-key model the active site of the enzyme is fixed and rigid, but in the induced-fit model the active site can change its shape slightly to accommodate the shape of the substrate.

16.11 a. Four factors that affect the rate of enzyme activity are temperature, pH, substrate concentration, and enzyme concentration.

b. Increase in temperature: Temperature increases the rate of a reaction, but if the temperature is high enough to denature the protein enzyme, the rate will decrease.
Increase in pH: Each enzyme has an optimum pH for action; if the pH departs from this point the reaction will slow down.
Increase in substrate concentration: Increased rate of reaction until maximum enzyme capacity is reached; after this no further increase.
Increase in enzyme concentration: Reaction rate increases.

Answers to Self-Test

The numbers in parentheses refer to sections in your textbook:
1. T (16.3) **2.** F; another (16.5) **3.** T (16.4) **4.** T (16.4) **5.** F; is the same as (16.5)
6. F; at one end and an amino group at the other end (16.5) **7.** T (16.1)
8. F; primary, secondary, tertiary, and quaternary (16.6) **9.** F; secondary (16.8)
10. F; secondary, tertiary, and quaternary (16.12) **11.** T (16.10) **12.** T (16.8)
13. T (16.14) **14.** F; function (16.14) **15.** F; apoenzyme (16.14) **16.** T (16.5) **17.** F; rigid (16.15)
18. T (16.14) **19.** F; nonprotein part (16.14) **20.** a (16.2) **21.** b (16.5) **22.** d (16.9) **23.** d (16.12)
24. c (16.11) **25.** a (16.13) **26.** a (16.14)

Nucleic Acids   Chapter 17

Chapter Overview

Nucleic acids are the molecules of heredity. Every inherited trait of every living organism is coded in these huge molecules. The complexity of their structure has been unraveled in fairly recent times, and this new knowledge of the transmission of genetic information has led to the exciting field of recombinant DNA technology.

In this chapter you will name and identify the structures of nucleotides and nucleic acids. You will write shorthand forms for nucleotide sequences in segments of DNA and RNA. You will identify the amino acid sequence coded by a given segment of DNA, and describe the processes of replication, transcription, and translation leading to protein synthesis. You will learn the basic ideas of recombinant DNA technology and genetic engineering.

Practice Exercises

17.1 **Nucleotides** (Sec. 17.2) are the structural units from which the polymeric **nucleic acids** (Sec. 17.1) are formed. A nucleotide is composed of a pentose sugar bonded to a phosphate group and a nitrogen-containing heterocyclic base. The identities of the sugars and bases differ in ribonucleic acid (RNA) and deoxyribonucleic acids (DNA).

Complete the table below identifying the pentoses and bases found in the nucleotides that make up RNA and DNA.

Nucleotide components	RNA	DNA
pentose		
purine bases		
pyrimidine bases		

17.2 The names of eight nucleotides, the monomer units making up RNA and DNA, are listed in Table 17.1 in your textbook. Both the names and the abbreviations of these names are commonly used. Complete the table below to acquaint yourself with the way nucleotides are named..

Name of nucleotide	Abbreviation	Base	Sugar
deoxyadenosine 5'-monophosphate			
	dTMP		
		guanine	ribose
	CMP		

17.3 The primary structure of a nucleic acid consists of the order in which the nucleotides. are linked together. Both RNA and DNA have an alternating sugar-phosphate backbone with the nitrogen-containing bases as side chain components.

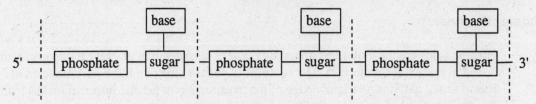

The end of the nucleotide chain that has a free phospate group attached to the 5' carbon is called the 5' end, and the end with a free hydroxyl group attached to the 3' carbon atom is the 3' end. The strand is read from the 5' end to the 3' end.

Draw the structural formula for the trinucleotide that forms between dAMP, dTMP, and dCMP, so that dAMP is the 5' end and to the left in your drawing, and dTMP is the 3' end and to the right in your drawing. Use a structural block diagram similar to the one above, but replace "base" with the specific name of the base, and replace "sugar" with the name of the sugar.

17.4 The structure of DNA is that of a double helix, in which two strands of DNA are coiled around each other in a spiral. The two strands are held together by hydrogen bonds between two pairs of **complementary bases** (Sec. 17.4), A–T and G–C. The relative amounts of these base pairs are constant for a given life form.

If, in a DNA molecule, the percentage of the base adenine is 20% of the total bases present, what would be the percentages of the bases thymine, cytosine, and guanine?

17.5 In the DNA double helix, the two complementary strands run in opposite directions, one in the 5' to 3' direction, the other 3' to 5'. Complete the following segment of a DNA double helix. Write symbols for the missing bases. Indicate the correct number of hydrogen bonds between the bases in each pair. (See Figure 17.8 in your textbook.)

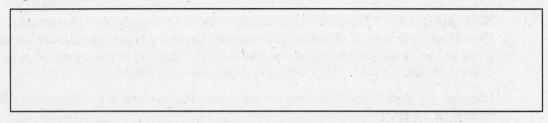

17.6 During **DNA replication** (Sec. 17.5), the DNA molecule makes an exact duplicate of itself. The two strands unwind, and free nucleotides line up along each strand, with complementary base pairs attracted to one another by hydrogen bonding. Polymerization of the new strand takes place. The daughter strand (3' to 5') is opposite in direction to the parent strand (5' to 3').

Write the sequence of bases for the replication of the DNA strand below:

5′ T–A–A–G–C–G–T–G–G 3′

17.7 There are four types of RNA molecules involved in the process of protein synthesis. Each type has a different function. Complete the table below on the different types of RNA.

Type of RNA	Abbreviation	Function
ribosomal RNA		
messenger RNA		
primary transcript RNA		
transfer RNA		

17.8 During **transcription** (Sec. 17.8), one strand of a DNA molecule acts as the template for the formation of a molecule of RNA. The nucleotides that line up next to the DNA strand have ribose as a sugar; the same bases are present except that uracil is substituted for thymine.

Using the new DNA strand formed in Practice Exercise 17.6, write the sequence of bases in the new RNA strand formed by transcription. Hint: Rewrite the answer in Practice Exercise 17.6, with 5′ on the left side, or starting point.

17.9 During **translation** (Sec. 17.6) RNA directs the synthesis of proteins; mRNA carries the code from DNA, and it is translated into the correct series of amino acids in the protein. A **codon** (Sec. 17.2) is a sequence of three nucleotides in an mRNA molecule that codes for a specific amino acid.

a. Complete the tables below with correct amino acid names and codons. The information can be obtained from Table 17.2 in your textbook.

Codon	Amino acid
UCA	
	asparagine
GAC	

Codon	Amino acid
	methionine
GAU	
	tryptophan

b. Are there any synonyms among the codons in the table in part a.?

c. Why is GTC not listed in Table 17.2 in your textbook as one of the codon sequences?

17.10   a.   Write a base sequence for mRNA that codes for the tripeptide Gly-Pro-Leu

         b.   Will there be only one answer? Explain.

17.11   An **anticodon** (Sec. 17.10) is a three-nucleotide sequence on tRNA that complements the
         mRNA sequence for the amino acid that bonds to that tRNA. Complete the table below for
         codons, their anticodons, and the amino acids they specify.

Codon	CAU		
Anticodon		GAG	
Amino acid			Trp

17.12   The two main processes of protein synthesis are transcription, in which DNA directs the
         synthesis of RNA molecules, and translation, in which RNA directs the synthesis of proteins.
         These two processes consist of various steps, reviewed in the table below. Tell what happens
         in each step and which molecules are involved.

Step	Process	Molecules involved
1. formation of ptRNA		
2. removal of introns		
3. mRNA to cytoplasm		
4. activation of tRNA		
5. initiation		
6, elongation		
7. termination		

Self-Test

True-false: Indicate whether the following statements are true or false. If the statement is false, give the word or phrase that may be substituted for the underlined portion to make the statement true.

1. The sugar unit found in DNA molecules is <u>ribose</u>.

2. In DNA the amount of adenine is equal to the amount of <u>guanine</u>.

3. DNA molecules are <u>the same</u> for individuals of the same species.

4. <u>Transfer RNA</u> molecules carry the genetic code from DNA to the ribosomes.

5. A codon is a series of <u>three</u> adjacent bases that carry the code for a specific amino acid.

6. A <u>gene</u> is an individual DNA molecule bound to a group of proteins.

7. The two strands of a DNA molecule are connected to each other by <u>hydrogen bonds</u> between the base units.

8. The process by which a DNA molecule forms an exact duplicate of itself is called <u>transcription</u>.

9. Primary transcript RNA is edited under the direction of enzymes and joined together to form <u>messenger RNA</u>.

10. Two different codons that specify the same amino acid are called <u>synonyms</u>.

11. Different species of organisms usually have <u>the same</u> genetic code for an amino acid.

12. A single mRNA molecule can serve as a codon sequence for the synthesis of <u>one protein molecule at a time</u>.

13. A complex of mRNA with several ribosomes proceeding along the ribosome molecule is called a <u>gene</u>.

Multiple choice:

14. Deoxyribose differs from ribose in that it has no –OH group on carbon number

    a. 1   b. 2   c. 3   d. 4   e. 5

15. The codon 5′ UGC 3′ would have as its anticodon:

    a. 5′ ACG 3′   b. 5′ UAG 3′   c. 5′ GAC 3′
    d. 5′ AUC 3′   e. none of these

16. Sections of DNA that carry noncoding base sequences are called:

    a. introns   b. exons   c. codons
    d. anticodons   e. none of these

17. Fifteen nucleotide units in a DNA molecule can contain the code for no more than:

    a. 3 amino acids   b. 5 amino acids   c. 10 amino acids
    d. 15 amino acids   e. none of these

18. Which of the following types of molecules does not carry information for protein synthesis?

    a. DNA   b. ribosomal RNA   c. messenger RNA
    d. transfer RNA   e. none of these

19   A sequence of three nucleotides in an mRNA molecule is a(n):

   a. exon   b. intron   c. codon
   d. anticodon   e. none of these

20. The intermediary molecules that deliver amino acids to the ribosomes are:

   a. ptRNA   b. tRNA   c. rRNA
   d. mRNA   e. none of these

21. The codon that initiates protein synthesis when it occurs as the first codon in an
     amino acid sequence is:

   a. GTA   b. UGA   c. GAC
   d. AUG   e. none of these

22. The process of inserting foreign DNA into a host cell is:

   a. translation   b. transformation   c. transcription
   d. transmission   e. none of these

23. Cells that have descended from a single cell and have identical DNA are called:

   a. mutagens   b. mutations   c. clones
   d. plasmids   e. none of these

Answers to Practice Exercises

17.1

Nucleotide components	RNA	DNA
pentose	ribose	deoxyribose
purine bases	adenine, guanine	adenine, guanine
pyrimidine bases	uracil, cytosine	cytosine, thymine

17.2

Name of nucleotide	Abbreviation	Base	Sugar
deoxyadenosine 5′-monophosphate	dAMP	adenine	deoxyribose
deoxythymidine 5′-monophosphate	dTMP	thymine	deoxyribose
guanosine 5′-monophosphate	GMP	guanine	ribose
cytidine 5′-monophosphate	CMP	cytosine	ribose

17.3

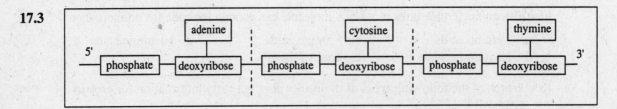

17.4   We know that %A = %T and %G = %C, since these bases are paired in DNA.

If %A = 20%, then %T = 20%. %A + %T = 40%,

so %G + %C = 60% and %G = %C = 30%

17.5   5′ T – C – A – C – A – G – T – A 3′
             ‖   ‖‖   ‖   ‖‖   ‖   ‖‖   ‖   ‖
        3′ A – G – T – G – T – C – A – T 5′

17.6   3′ A–T–T–C–G–C–A–C–C 5′   or   5′ C–C–A–C–G–C–T–T–A 3′

17.7

Type of RNA	Abbreviation	Function
ribosomal RNA	rRNA	site for protein synthesis
messenger RNA	mRNA	carries genetic information from DNA to ribosomes
primary transcript RNA	ptRNA	material from which mRNA is made
transfer RNA	tRNA	delivers individual amino acids to ribosomes for protein synthesis

17.8   5′ C–C–A–C–G–C–T–T–A 3′   DNA

        3′ G–G–U–G–C–G–A–A–U 5′   RNA

17.9

Codon	Amino acid
UCA	serine
AAU and AAC	asparagine
GAC	aspartic acid

Codon	Amino acid
AUG	methionine
GAU	aspartic acid
UGG	tryptophan

    b.   Yes; GAU and GAC both code for aspartic acid, and AAU and AAC both code for asparagine.

    c.   GTC is not listed as a codon because RNA does not contain thymine.

17.10 a.   5′ G–G–U–C–C–C–C–U–U 3′ is one possible answer.

         b.   No, because there is more than one codon for most amino acids.

17.11

Codon	CAU	CUC	UGG
Anticodon	GUA	GAG	ACC
Amino acid	His	Leu	Trp

17.12

Step	Process	Molecules involved*
1. Formation of ptRNA	DNA unwinds, acts as template for ptRNA formation	DNA, ptRNA, nucleotides
2. Removal of introns	ptRNA strand cut, introns removed, mRNA bonds form	ptRNA, mRNA
3. mRNA to cytoplasm	mRNA moves out of the nucleus, into the cytoplasm	mRNA
4. Activation of tRNA	tRNA attaches to an amino acid and becomes energized	tRNA, amino acid, ATP
5. Initiation	mRNA attaches to ribosome, tRNA with amino acid moves to first codon	mRNA, tRNA with attached amino acid, rRNA
6. Elongation	more tRNA molecules move to next codons, polypeptide chain transfers to each new tRNA	mRNA, tRNA with attached amino acid, rRNA
7. Termination	stop codon appears on mRNA, peptide chain (protein) is cleaved from tRNA	mRNA, tRNA, protein

*Enzymes are also involved at each step of the process.

Answers to Self-Test

The numbers in parentheses refer to sections in your textbook:
1. F; deoxyribose (17.2) **2.** F; thymine (17.4) **3.** F; different (17.4) **4.** F; messenger RNA (17.7)
5. T (17.9) **6.** F; chromosome (17.5) **7.** T (17.4) **8.** T (17.6) **9.** T (17.7) **10.** T (17.9) **11.** T (17.9)
12. F; many protein molecules at a time (17.7) **13.** F; polysome (17.11) **14.** b (17.2) **15.** a (17.10)
16. a (17.8) **17.** b (17.9) **18.** b (17.7) **19.** c (17.9) **20.** b (17.7) **21.** d (17.11) **22.** b (17.12)
23. c (17.12)

Metabolism   Chapter 18

Chapter Overview

The most important job of the body's cells is the production of energy to be utilized in carrying out the complex processes of life. The production and use of energy by living organisms involves an important intermediate called ATP.

In this chapter you will study the formation of acetyl CoA from the products of the digestion of food, and the further oxidation of acetyl CoA during the individual steps of the citric acid cycle. You will study the function and processes of the electron transport chain, and the important role of ATP in energy transfer.

The complete oxidation of glucose begins with glycolysis. You will study the reactions of glycolysis to produce pyruvate, and the pathways of pyruvate under aerobic and anaerobic conditions,

Practice Exercises

18.1   During the processes of **metabolism** (Sec. 18.1) food is converted into energy. Metabolic reactions take place in various locations within the cell. A eukaryotic cell, in which DNA is inside a membrane enclosed nucleus, is shown below. Write the letters from the diagram next to the appropriate descriptions below the diagram.

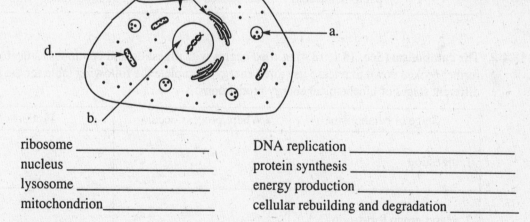

ribosome _____	DNA replication _____
nucleus _____	protein synthesis _____
lysosome _____	energy production _____
mitochondrion_____	cellular rebuilding and degradation _____

18.2   During the metabolic reactions that convert food to energy, several compounds function as key intermediates. To show how these compounds are structurally related to one another, complete the block diagrams below. Use information from Sec. 18.3 of your textbook.

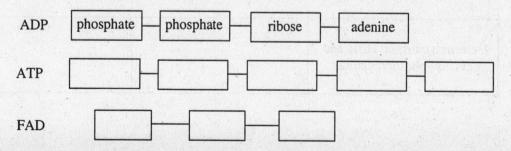

NAD⁺
[box] — [box] — [box]

Coenzyme A
[box] — [box] — [box]

18.3   The matching exercise below will help to clarify the structures and actions of key intermediates in metabolic reactions.

Answer	Intermediates (or their components)	Explanation or definition
	nicotinamide	1. oxidized form of nicotinamide adenine dinucleotide
	NADH	2. vitamin that is a part of FAD
	riboflavin	3. transfers acetyl groups
	FAD	4. vitamin that is a part of coenzyme A
	coenzyme A	5. reduced form of nictoninamide adenine dinucleotide
	$FADH_2$	6. oxidized form of flavin adenine dinucleotide
	NAD^+	7. vitamin that is a part of NAD^+
	pantothenic acid	8. reduced form of flavin adenine dinucleotide

18.4   The **catabolism** (Sec. 18.1, 18.4) of food begins with digestion and continues as the food is further broken down to release its stored energy. Complete the following table for the four different stages of biochemical energy production.

Stage of catabolism	Where process occurs	Products
1. digestion		
2. acetyl group formation		
3. citric acid cycle		
4. electron transport chain and oxidative phosphorylation		

18.5 The **citric acid cycle** (Sec. 18.5) is the series of reactions in which the acetyl group of acetyl CoA is oxidized to CO_2, and $FADH_2$ and NADH are produced. Complete the following table summarizing the steps of the citric acid cycle. Use the discussion and equations in section 18.5 of your textbook.

Step	Type of reaction	Final product(s)	Energy transfer intermediates
1.	condensation		none
2.		isocitrate	
3.			NADH
4.			
5.			
6.			
7.			
8.			

18.6 Four of the steps summarized in Practice Exercise 18.5 involve oxidation and reduction. For each of these steps, give the name and/or symbol for the substance that was oxidized and for the one that was reduced.

Step	Substance oxidized	Substance reduced
3.		
4.	α-ketoglutarate	
6.		FAD (to $FADH_2$)
8.		

18.7 The **electron transport chain** (Sec. 18.6) is a series of reactions in which electrons and hydrogen ions from NADH and $FADH_2$ are passed to intermediate carriers and ultimately react with molecular oxygen to produce water.

Use the summary of the electron transport chain given in section 18.6 in your textbook to complete the following table.

Step	Substance oxidized	Substance reduced
1.		
2.		
3.		
4.		
5.		
6.		
7.		
8.		
final step		

18.8 **Oxidative phophorylation** (Sec. 18.7) is the process by which ATP is synthesized from ADP using energy released in the electron transport chain.

Using Practice Exercise 18.5, summarize the number of ATP molecules produced in one turn of the citric acid cycle (steps 1 through 8). Remember: Each NADH from the citric acid cycle produces 2.5 ATP molecules, each $FADH_2$ produces 1.5 ATP molecules, and each GTP produces 1 ATP molecule.

Step	Energy-rich compound formed	Number of ATP's produced
3.		
4,		
5.		
6.		
8.		
	Total:	

18.9 **Glycolysis** (Sec. 18.9) is the metabolic pathway by which glucose is converted into two molecules of pyruvate. Using the steps of glycolysis shown in Figure 18.15 and discussed in section 18.9 of your textbook, give the number(s) of the reaction steps for each of the following:

a. Where are ATPs produced? _____

b. Where are ATPs used?   _____

c. Where is NAD^+ reduced? _____

d. Where is the carbon chain split? _____

e. Where is a ketone isomerized to an aldehyde? _____

f. Where are phosphate groups added to sugar molecules? _____

g. Where are phosphate groups removed from sugar molecules? _____

h. Where is water lost? _____

18.10 The pyruvate produced by glycolysis reacts further in several different ways according to the conditions and the type of organism. Complete the following table on the fates of pyruvate produced by glycolysis.

Conditions	Name of process	Name of product	Number of NADH used or produced
1. aerobic			
2. anaerobic (humans)			
3. anaerobic (yeasts)			

18.11 You have studied the many reactions involved in the complete oxidation of one glucose molecule. Below is a summary of the reactions in each of the four stages of this oxidation. Because of the complexity of the processes, the "equations" are unbalanced. To get an overview of the process, work with them as you would with ordinary balanced equations.

Add the equations together by crossing out molecules that occur on both the reactant side of one equation and the product side of another. Write the molecules that remain as the final net equation for the entire process. Balance each type of atom in the final equation. Sum up the ATPs produced.

Glycolysis

$$glucose \longrightarrow 2\ pyruvate$$

$$2\ NAD^+ \longrightarrow 2\ NADH_{cytochrome}$$

$$\Big\} \quad 2\ ATP$$

Oxidation of Pyruvate

$$2\ pyruvate \longrightarrow 2\ acetyl\ CoA$$

$$2\ NAD^+ \longrightarrow 2\ NADH$$

$$\Big\} \quad 0\ ATP$$

Citric Acid Cycle

$$2\ acetyl\ CoA \longrightarrow CO_2 + H_2O$$

$$2\ FAD \longrightarrow 2\ FADH_2$$

$$6\ NAD^+ \longrightarrow 6\ NADH$$

$$\Big\} \quad 2\ ATP$$

Electron Transport Chain

$$2\ NADH_{cytochrome} \longrightarrow 2\ NAD^+$$

$$2\ FADH_2 \longrightarrow 2\ FAD$$

$$8\ NADH \longrightarrow 8\ NAD^+$$

$$O_2 \longrightarrow H_2O$$

$$\Big\} \quad 26\ ATP$$

- -

$$C_6H_{12}O_6 + O_2 \longrightarrow CO_2 + H_2O \qquad total\ ATP =$$
(glucose)

Self-Test

True-false: Indicate whether the following statements are true or false. If the statement is false, give the word or phrase that may be substituted for the underlined portion to make the statement true.

1. Catabolic reactions usually <u>release</u> energy.
2. Bacterial cells do not contain a nucleus and so are classified as <u>eukaryotic</u>.
3. <u>Ribosomes</u> are organelles that have a central role in the production of energy.
4. The active portion of the FAD molecule is the <u>flavin</u> subunit.
5. <u>NAD^+</u> is the oxidized form of nicotine adenine dinucleotide.
6. The citric acid cycle occurs in the <u>cytoplasm</u> of cells.
7. In the first step of the citric acid cycle, oxaloacetate reacts with <u>glucose</u>.
8. The electron transport chain receives electrons and hydrogen ions from <u>NAD^+ and FAD</u>.
9. The electrons that pass through the electron transport chain <u>gain energy</u> in each transfer along the chain.
10. Cytochromes contain <u>iron</u> atoms that are reversibly oxidized and reduced.
11. ATP is produced from ADP using energy <u>released</u> in the electron transport chain.
12. The process by which glucose is degraded to ethanol is called <u>fermentation</u>.
13. In human metabolism, under anaerobic conditions, pyruvate is reduced to <u>acetyl CoA</u>.
14. The complete oxidation of glucose in skeletal muscle and nerve cells yields <u>30</u> ATP molecules per glucose molecule.

Multiple choice:

15. The citric acid cycle takes place in which of the following parts of the cell?

    a. mitochondrion   b. nucleus   c. ribosome
    d. lysosome   e. none of these

16. Which of these molecules is **not** part of the electron transport chain?

    a. coenzyme Q   b. acetyl CoA   c. cytochrome b
    d. ATP   e. none of these

17. The final acceptor of electrons in the electron transport chain is:

    a. NAD^+ b. water   c. oxygen
    d. FAD   e. none of these

Use the discussion of the citric acid cycle in Section 18.5 of your textbook as you answer the following questions:

18. Which of these steps in the citric acid cycle is involved with the removal of a CO_2 molecule?

    a. 1   b. 3   c. 5   d. 7   e. 8

19. Which step in the citric acid cycle involves the removal of hydrogen atoms from the carbon chain to form an alkene?

    a. 1   b. 4   c. 6   d. 7   e. none of these

20. Which of these steps in the citric acid cycle involves the oxidation of a secondary alcohol to a ketone?

   a. 2   b. 4   c. 6   d. 7   e. 8

Refer to the steps of glycolysis discussed in Section 18.9 of your textbook as you answer the next three questions:

21. In glycolysis, which steps involve the removal of a phosphate group from ATP?

   a. 1 and 3   b. 2 and 5   c. 4 and 3   d. 1 and 5   e. 5 and 7

22. In glycolysis, the formation of ATP from ADP occurs during which step?

   a. 1   b. 4   c. 5   d. 6   e. 7

Answers to Practice Exercises

18.1

ribosome —————— c.	DNA replication —————————— b.
nucleus ————— b.	protein synthesis ————————— c.
lysosome ————— a.	energy production ————————— d.
mitochondrion ——— d.	cellular rebuilding and degradation — a.

18.2

ADP   [phosphate]—[phosphate]—[ribose]—[adenine]

ATP   [phosphate]—[phosphate]—[phosphate]—[ribose]—[adenine]

FAD   [flavin]—[ribitol]—[ADP]

NAD^+ [nicotinamide]—[ribose]—[ADP]

Coenzyme A   [2-aminoethanethiol]—[pantothenic acid]—[phosphorylated ADP]

18.3

Answer	Intermediates (or their components)	Explanation or definition
7.	nicotinamide	1. oxidized form of nicotinamide adenine dinucleotide
5.	NADH	2. vitamin that is a part of FAD
2.	riboflavin	3. transfers acetyl groups
6.	FAD	4. vitamin that is a part of coenzyme A
3.	coenzyme A	5. reduced form of nictoninamide adenine dinucleotide
8.	$FADH_2$	6. oxidized form of flavin adenine dinucleotide
1.	NAD^+	7. vitamin that is a part of NAD^+
4.	pantothenic acid	8. reduced form of flavin adenine dinucleotide

18.4

Stage of catabolism	Where process occurs	Major products
1. digestion	mouth, stomach, small intestine	glucose, fatty acids and glycerol, amino acids
2. acetyl group formation	cytoplasm of cell and inside mitochondria	acetyl CoA
3. citric acid cycle	inside mitochondria	CO_2, water, NADH, and $FADH_2$
4. electron transport chain and oxidative phosphorylation	inside mitochondria	water and ATP

18.5

Step	Type of reaction	Final product(s)	Energy transfer intermediates
1.	condensation	citrate, free coenzyme A	none
2.	isomerization	isocitrate	none
3.	oxidation, decarboxylation	α-ketoglutarate, CO_2	NADH
4.	oxidation, decarboxylation	succinyl CoA, CO_2	NADH
5.	phosphorylation	succinate, free coenzyme A	GTP
6.	oxidation (dehydrogenation)	fumarate	$FADH_2$
7.	hydration	L-malate	none
8.	oxidation (dehydrogenation)	oxaloacetate	NADH

18.6

Step	Substance oxidized	Substance reduced
3.	isocitrate	NAD^+ (to NADH)
4.	α-ketoglutarate	NAD^+ (to NADH)
6.	succinate	FAD (to $FADH_2$)
8.	L-malate	NAD^+ (to NADH)

18.7

Step	Substance oxidized	Substance reduced
1.	NADH	FMN
2.	$FMNH_2$	Fe(III)
3.	Fe(II) (or $FADH_2$)	CoQ
4.	$CoQH_2$	Fe(III) cytochrome b
5.	Fe(II) cytochrome b	Fe(III) cytochrome c_1
6.	Fe(II) cytochrome c_1	Fe(III) cytochrome c
7.	Fe(II) cytochrome c	Fe(III) cytochrome a
8.	Fe(II) cytochrome a	Fe(III) cytochrome a_3
final step	Fe(II) cytochrome a_3	O_2

18.8

Step	Energy-rich compound formed	Number of ATP's produced
3.	NADH	2.5
4.	NADH	2.5
5.	GTP	1
6.	$FADH_2$	1.5
8.	NADH	2.5
	Total:	10

18.9   a.   ATPs are produced in steps 7 and 10.

b.   ATPs are used in steps 1 and 3.

c. NAD^+ is reduced in step 6.

d.   The carbon chain is split in step 4.

e.   A ketone is isomerized to an aldehyde in step 5.

f.   Phosphate groups are added to sugar molecules in steps 1, 3, and 6.

g.   Phosphate groups are removed from sugar molecules in steps 7 and 10.

h.   Water is lost in step 9.

18.10

Conditions	Name of process	Product	Number of NADH
1. aerobic	oxidation	acetyl	1 produced
2. anaerobic (humans)	reduction	lactate	1 used
3. anaerobic (yeasts)	fermentation	ethanol	1 used

18.11

Glycolysis

$$glucose \longrightarrow 2\ pyruvate$$

$$2\ NAD^+ \longrightarrow 2\ NADH_{cytochrome}$$

} 2 ATP

Oxidation of Pyruvate

$$2\ pyruvate \longrightarrow 2\ acetyl\ CoA$$

$$2\ NAD^+ \longrightarrow 2\ NADH$$

} 0 ATP

Citric Acid Cycle

$$2\ acetyl\ CoA \longrightarrow CO_2 + H_2O$$

$$2\ FAD \longrightarrow 2\ FADH_2$$

$$6\ NAD^+ \longrightarrow 6\ NADH$$

} 2 ATP

Electron Transport Chain

$$2\ NADH_{cytochrome} \longrightarrow 2\ NAD^+$$

$$2\ FADH_2 \longrightarrow 2\ FAD$$

$$8\ NADH \longrightarrow 8\ NAD^+$$

$$O_2 \longrightarrow H_2O$$

} 26 ATP

===

$$C_6H_{12}O_6 + 6\ O_2 \longrightarrow 6\ CO_2 + 6\ H_2O \qquad total\ ATP = 30\ ATP$$
(glucose)

Answers to Self-Test

The numbers in parentheses refer to sections in your textbook:
1. T (18.1) **2.** F; prokaryotic (18.2) **3.** F; mitochondria (18.2) **4.** T (18.3) **5.** T (18.3)
6. F; mitochondria (18.5) **7.** F; acetyl CoA (18.5) **8.** F; NADH and FADH$_2$ (18.6)
9. F; lose energy (18.6) **10.** T (18.6) **11.** T (18.6) **12.** T (18.10) **13.** F; lactate (18.10)
14. T (18.11) **15.** a (18.2) **16.** b (18.6) **17.** c (18.6) **18.** b (18.5) **19.** c (18.5) **20.** e (18.5)
21. a (18.9) **22.** e (18.9)

Answer to Questions and Problems

1.1   a. matter   b. matter   c. energy   d. energy   e. matter   f. matter

1.3   a. shape (indefinite for liquids and definite for solids)
       b. volume (indefinite for gases and definite for liquids)

1.5   a. No, it has a definite shape.   b. No, it has an indefinite volume.
       c. yes   d. yes

1.7   a. physical   b. physical   c. chemical   d. chemical

1.9   a. chemical   b. physical   c. chemical   d. physical

1.11   a. heterogeneous mixture   b. homogeneous mixture
       c. pure substance   d. heterogeneous mixture

1.13   a. true
       b. false, compounds involve a chemical combination rather than physical combination
       c. false, neither compounds nor elements can have a variable composition
       d. true

1.15   a. A, cannot classify; B, cannot classify; C, compound
       b. D, compound; E, cannot classify; F, cannot classify

1.17   a. true
       b. false, hydrogen is the most abundant element in the human body
       c. true
       d. true

1.19   a. yes
       b. no, all start with the first letter of the element's English name
       c. no, cobalt and chromium have symbols that start with C
       d. yes

1.21   a. heteroatomic, diatomic, compound   b. heteroatomic, triatomic, compound
       c. homoatomic, diatomic, element   d. heteroatomic, triatomic, compound

1.23   a. 3 elements, 6 atoms   b. 4 elements, 10 atoms   c. 3 elements, 6 atoms
       d. 2 elements, 14 atoms   e. 3 elements, 5 atoms   f. 3 elements, 9 atoms

1.25   a. XQ b. XQ_2 c. Q_3 d. X_2Q

1.27   a. element   b. mixture   c. mixture   d. compound

1.29   a. an element and a compound   b. an element and a compound
       c. two elements   d. a single pure substance

1.31   a. same, both 4   b. more, 6 and 5   c. same, both 5   d. fewer, 13 and 15

1.33   a. 2 (N_2, NH_3)   b. 4 (N, H, C, Cl)
       c. 110, 5(2 + 6 + 4 + 5 + 5)   d. 56, 4(4 + 3 + 4 + 3)

1.35   true

1.37   false, elements are known for all elements with atomic numbers between 1 and 112

1.39   false, compounds result from the chemical combination of elements

1.41   true

1.43   false, "melts at 73°C" is a physical property

1.45   true

1.47   false, 20 atoms are present per formula unit

1.49   b

1.51   c

1.53   d

1.55   c

1.57   d

Answers to Questions and Problems

2.1   mass, volume, length, time, temperature, pressure, and concentration

2.3   a. kilo-   b. milli-   c. micro-   d. deci-

2.5   a. nanogram, milligram, centigram   b. kilometer, megameter, gigameter
       c. picoliter, microliter, deciliter   d. microgram, milligram, kilogram

2.7   a. inexact   b. exact   c. exact   d. inexact

2.9   Because only one estimated digit may be recorded as part of a measured value.

2.11   a. the 1   b. the 0   c. the 5   d. the 3

2.13   a. yes (4 and 4)   b. no (4 and 3)   c. yes (2 and 2)   d. yes (4 and 4)

2.15   a. two   b. two   c. two   d. two

2.17   a. 162   b. 9.3   c. 1261   d. 20.0

2.19   a. 1.0×10^{-3} b. 1.0×10^{3} c. 6.3×10^{4} d. 6.3×10^{-4}

2.21   a. $1 \text{ kg}/10^{3} \text{ g}$ and $10^{3} \text{ g}/1 \text{ kg}$ b. $1 \text{ nm}/10^{-9} \text{ m}$ and $10^{-9} \text{ m}/1 \text{ nm}$
       c. $1 \text{ mL}/10^{-3} \text{ L}$ and $10^{-3} \text{ L}/1 \text{ mL}$ d. $1 \text{ cg}/10^{-2} \text{ g}$ and $10^{-2} \text{ g}/1 \text{ cg}$

2.23   a. 60 seconds = 1 minute   b. 12 in. = 1 ft
       c. 2.54 cm = 1.00 in.   d. 454 g = 1.00 lb

2.25   a. $6.4 \text{ g} \times \dfrac{1.00 \text{ lb}}{454 \text{ g}} = 0.014 \text{ lb}$ b. $6.4 \text{ g} \times \dfrac{454 \text{ g}}{1.00 \text{ lb}} = 2900 \text{ g}$

       c. $53 \text{ cm} \times \dfrac{1.00 \text{ in.}}{2.54 \text{ cm}} = 21 \text{ in.}$ d. $3.5 \text{ qt} \times \dfrac{0.946 \text{ L}}{1.00 \text{ qt}} = 3.3 \text{ L}$

2.27 $2500 \text{ mL} \times \dfrac{10^{-3} \text{ L}}{1 \text{ mL}} = 2.5 \text{ L}$

2.29 $\dfrac{524.5 \text{ g}}{38.72 \text{ cm}^{3}} = 13.55 \text{ g} / \text{cm}^{3}$

2.31 $15 \text{ cm}^{3} \times \dfrac{8.90 \text{ g}}{1 \text{ cm}^{3}} = 130 \text{ g}$

2.33 $9/5(35°) + 32° = 95°F$

2.35 a. 2.0 calories; a calorie is 4.184 times larger than a joule
      b. 1.0 kilocalorie; 1.0 kilocalorie = 1.0 x 10^3 calories
      c. 100 Calories; 1 Calorie = 1000 calories
      d. 1000 kilocalories; 2.3 Calories = 2.3 kilocalories

2.37 a. 3.00 x 10^{-3} b. 9.4 x 10^5 c. 2.35 x 10^1 d. 4.50000 x 10^8

2.39 a. four significant figures   b. four significant figures
      c. three significant figures   d. exact

2.41 a. two   b. two   c. three   d. three

2.43 $-10°C$; $10°F = -12°C$

2.45 An exact number cannot have digits to the right of the decimal point.

2.47 true

2.49 false, the three leading zeros are not significant

2.51 false, gram is the base unit of mass in the metric system

2.53 false, exact numbers have no uncertainty

2.55 true

2.57 true

2.59 true

2.61 c

2.63 c

2.65 d

2.67 b

2.69 c

Answers to Questions and Problems

3.1   a. electron   b. neutron   c. proton   d. proton

3.3   Protons present in a nucleus give it its positive charge.

3.5   a. true
      b. false, both protons and neutrons are present in nuclei
      c. false, nuclear volume is very small compared to atomic volume
      d. false, electrons are not found in a nucleus

3.7   a. 8 protons, 8 neutrons, 8 electrons   b. 8 protons, 10 neutrons, 8 electrons
      c. 20 protons, 24 neutrons, 20 electrons   d. 100 protons, 157 neutrons, 100 electrons

3.9   a. number of protons or number of electrons   b. number of nucleons
      c. number of neutrons   d. number of subatomic particles

3.11 atomic number = 12; mass number = 24

3.13 a. 40 protons, 40 electrons, 50 neutrons   b. 40 protons, 40 electrons, 52 neutrons
      c. 40 protons, 40 electrons, 55 neutrons   d. 40 protons, 40 electrons, 56 neutrons

3.15 $^{12}_{6}C$, $^{13}_{6}C$, $^{14}_{6}C$

3.17 (0.9258 x 7.02 amu) + (0.742 x 6.01 amu) = 6.95 amu

3.19 a. Ca   b. Mo   c. Li   d. Sn

3.21 a. 6   b. 28.09 amu   c. 39   d. 9.01 amu

3.23 a. no (both are nonmetals)   b. no (Al is a metal, and Si is a nonmetal)
      c. yes   d. yes

3.25 a. metallic   b. nonmetallic   c. metallic   d. nonmetallic

3.27 a. 1 electron   b. 2 electrons   c. 3 electrons   d. 4 electrons

3.29 a. 2 electrons   b. 2 electrons   c. 2 electrons   d. 2 electrons

3.31 a. true
      b. true
      c. false, s subshells accommodate 2 electrons, p subshells 6 electrons, etc.
      d. true

3.33 a. oxygen   b. neon   c. aluminum   d. calcium

3.35   a. true
        b. true
        c. true
        d. false, there are 15 electrons present and element 15 is phosphorus

3.37   a. *s* area   b. *d* area   c. *f* area   d. *p* area

3.39   a. representative element   b. noble-gas element   c. transition element
        d. inner-transition element

3.41   spontaneous emission of radiation

3.43   for low atomic numbers, the neutron/proton ratio is about 1; for high atomic numbers, the
        neutron/proton ratio is about 1.5

3.45   a. 12 hr = 2 half-lives; $(1/2)^2 = 1/4$ b. 36 hr = 6 half-lives; $(1/2)^6 = 1/64$
        c. $(1/2)^3 = 1/8$ d. $(1/2)^6 = 1/64$

3.47   112 yr = 4 half-lives; $(1/2)^4 = 1/16$ undecayed; 15/16 has decayed; 15/16 x 4.00 g = 3.75 g

3.49   a. + 2, 4 amu   b. –1, 00055 amu   c. 0, 0 amu

3.51   They are identical.

3.53   a. $^{10}_{4}Be \rightarrow \,^{0}_{-1}\beta + \,^{10}_{5}B$ b. $^{77}_{32}Ge \rightarrow \,^{0}_{-1}\beta + \,^{77}_{33}As$

        c. $^{60}_{26}Fe \rightarrow \,^{0}_{-1}\beta + \,^{60}_{27}Co$ d. $^{25}_{11}Na \rightarrow \,^{0}_{-1}\beta + \,^{25}_{12}Mg$

3.55   no change in mass number; increase of 1 unit in atomic number

3.57   a. $^{190}_{78}Pt \rightarrow \,^{186}_{76}Os + \,^{4}_{2}\alpha$; alpha decay

        b. $^{19}_{8}O \rightarrow \,^{19}_{9}F + \,^{0}_{-1}\beta$; beta decay

3.59   Alpha does not penetrate; beta penetrates outer layers; gamma completely penetrates.

3.61   alpha, 6 cm with 40,000 collisions; beta, 1000 cm with 2000 collisions

3.63   so that they remain in the body only a short period of time

3.65   yttrium-90, implanted in the body; cobalt-60, external exposure

3.67   a. $^{44}_{20}Ca$ b. $^{9}_{4}Be$ c. $^{110}_{47}Ag$ d. $^{9}_{4}Be$

3.69   a. not isotopes   b. not isotopes   c. isotopes   d. isotopes

3.71   All fluorine atoms are identical (only one naturally occurring fluorine isotope exists); more
        than one isotope exists for iron (55.847 amu is the average mass for all isotopes).

3.73   a. $1s^2 2s^2 2p^1$ b. $1s^2 2s^2 2p^6 3s^2 3p^1$ c. $1s^2 2s^1$ d. $1s^2 2s^2$

3.75   a. boron   b. scandium   c. lithium   d. phosphorus

3.77   a. 1/8 remains undecayed; $1/8 = (1/2)^3$, 3 half-lives; 3 x 18 hr = 54 hr
        b. 1/32 remains undecayed; $1/32 = (1/2)^5$, 5 half-lives; 5 x 18 hr = 90 hr
        c. 1/64 remains undecayed; $1/64 = (1/2)^6$, 6 half-lives; 6 x 18 hr = 108 hr
        d. 1/128 remains undecayed; $1/128 = (1/2)^7$, 7 half-lives; 7 x 18 hr = 126 hr

3.79   false, a nucleon is any subatomic particle found in the nucleus

3.81   true

3.83   true

3.85   false, electron configurations group electrons according to subshells rather than orbitals

3.87   true

3.89   false, in the second half-life 1/4 (1/2 x 1/2) of the original atoms undergo decay

3.91   true

3.93   c

3.95   c

3.97   b

3.99   b

3.101   b

Answers to Questions and Problems

4.1   electron

4.3   Ionic compounds have high melting points and are good conductors of electricity in solution and the molten state; molecular compounds have lower melting points than ionic compounds and do not conduct electricity.

4.5   a. IA, one   b. VIIIA, eight   c. IIA, two   d. VIIA, seven

4.7   a. IIA, $Mg\cdot$ b. IA, $K\cdot$ c. VA, $\cdot\overset{\cdot}{\underset{\cdot}{P}}\cdot$ d. VIIIA, $\vdots\overset{\cdot\cdot}{\underset{\cdot\cdot}{Kr}}\vdots$

4.9   Elements with the same number of valence electrons will be in the same periodic table group.
         a. Be and Mg   b. N and P   c. O and S   d. Na and K

4.11   They are the most unreactive of all elements.

4.13   He has only two valence electrons.

4.15   a. O^{2-} b. Mg^{2+} c. F^- d. Al^{3+}

4.17   a. 15 protons, 18 electrons   b. 16 protons, 18 electrons
         c. 12 protons,10 electrons   d. 3 protons, 2 electrons

4.19   a. four electrons   b. three electrons   c. two electrons   d. one electron

4.21   a. +2   b. –1   c. –3   d. +3

4.23   a. IIA   b. VIA   c. VA   d. IA

4.25   It becomes a negative ion through electron gain.

4.27   a. $Be \rightleftarrows \overset{\cdot\cdot}{\underset{\cdot\cdot}{O}}\vdots$ b. $Mg \rightleftarrows \overset{\cdot}{\underset{\cdot\cdot}{S}}\vdots$ c. $\begin{matrix} K\cdot \\ K\cdot \\ K\cdot \end{matrix} \to \overset{\cdot}{\underset{\cdot}{N}}\vdots$ d. $Ca \rightrightarrows \begin{matrix} \vdots\overset{\cdot\cdot}{\underset{\cdot\cdot}{F}}\vdots \\ \vdots\overset{\cdot\cdot}{\underset{\cdot\cdot}{F}}\vdots \end{matrix}$

4.29   a. MgF_2 b. BeF_2 c. LiF   d. AlF_3

4.31   The positive ion always appears first in the chemical formula.

4.33   The formula unit is the smallest whole number repeating ratio of ions present in the ionic compound.

4.35   Binary ionic compounds contain a metal and a nonmetal.
         a. yes   b. yes   c. no   d. yes

4.37 Oxygen is a –2 in each case.
    a. $2(Ag) + (-2) = 0;\ Ag = +1$
    b. $Cu + (-2) = 0;\ Cu = +2$
    c. $Sn + 2(-2) = 0;\ Sn = +4$
    d. $Sn + (-2) = 0;\ Sn = +2$

4.39   a. gold(I) chloride   b. potassium chloride c. silver chloride   d. copper(II) chloride

4.41   a. CoO b. Co_2O_3 c. SnI_4 d. Pb_3N_2

4.43   ionic, electron transfer; covalent, electron sharing

4.45   a. :Br:Br: b. H:I: c. :I:Br: d. :Br:F:

4.47   a. H_2O b. CBr_4 c. PI_3 d. SiH_4

4.49   These are three terms that designate the same thing.

4.51   a. 0, 0, 1   b. 1, 0, 1   c. 0, 2, 0   d. 1, 0, 0

4.53   a.   b.

    c. (see structure)   d. (see structure)

4.55   a. H:P:H with H below   b. :Cl:P:Cl: with :Cl: below   c. :Br:Si:Br: with :Br: above and below   d. :F:O:F:

4.57   a. :Cl:C:Cl: with :O: below   b. :F:N::N:F: c. H:C::C:H with H below each C   d. H:C:::C:H

4.59   a. linear   b. angular   c. angular   d. angular

4.61   The shape does not change; it remains tetrahedral.

4.63   0.5 units

4.65   a. $\overset{\delta^+\ \ \delta^-}{B-N}$ b. $\overset{\delta^+\ \ \delta^-}{Cl-F}$ c. $\overset{\delta^-\ \ \delta^+}{N-C}$ d. $\overset{\delta^-\ \ \delta^+}{F-O}$

4.67   a. zero difference   b. greater than zero, less than 2.0   c. 2.0 or greater

4.69   a. nonpolar   b. polar   c. polar   d. polar

4.71   a. nonpolar   b. polar   c. nonpolar   d. polar

4.73   a. dichlorine monoxide   b. carbon monoxide
       c. phosphorus triiodide   d. hydrogen iodide

4.75   a. H_2O_2 b. NH_3 c. CH_4 d. N_2H_4

4.77   a. SO_4^{2-} and SO_3^{2-} b. NO_3^- and NO_2^-
       c. CO_3^{2-} and HCO_3^- d. HPO_4^{2-} and $H_2PO_4^-$

4.79   a. $Fe(OH)_3$ b. $Be(NO_3)_2$ c. $(NH_4)_2S$ d. $(NH_4)_3PO_4$

4.81   a. copper(I) phosphate   b. iron(III) nitrate
       c. iron(II) sulfate   d. gold(I) cyanide

4.83   a. C: $1s^2 2s^2 2p^2$ b. F: $1s^2 2s^2 2p^5$ c. Mg: $1s^2 2s^2 2p^6 3s^2$ d. P: $1s^2 2s^2 2p^6 3s^2 3p^3$

4.85   a. XZ_2 b. X_2Z c. XZ   d. ZX

4.87   a. ionic   b. molecular   c. ionic   d. molecular

4.89   a. beryllium chloride   b. nitrogen trichloride
       c. aluminum chloride   d. iron(III) chloride

4.91   a.

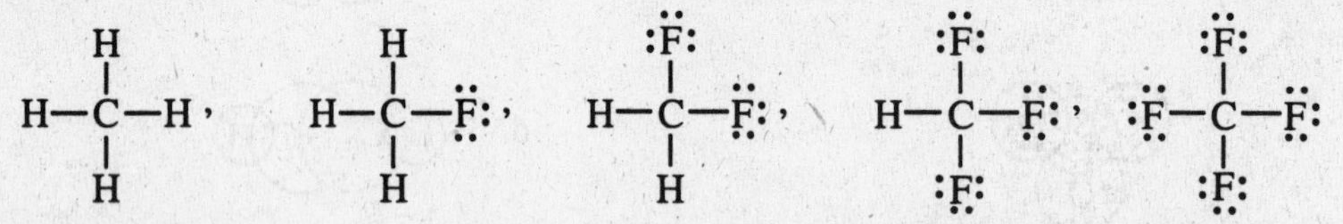

       b. All are tetrahedral.   c. nonpolar, polar, polar, polar, nonpolar

4.93   a. polar covalent, ionic, ionic, polar covalent
       b. BA, CA, DB, DA
       c. BA, CA, DB, DA

4.95   true

4.97   false, it is electron loss or gain rather than proton loss or gain

4.99   true

4.101 true

4.103 true

4.105 true

4.107 true

4.109 c

4.111 a

4.113 c

4.115 d

4.117 d

Chemical Calculations: Formula Masses, Moles, and Chemical Equations   Chapter 5

Answers to Questions and Problems

5.1   a. (12.01 amu) + 4(1.01 amu) = 16.05 amu
      b. (32.07 amu) + 3(16.00 amu) = 80.07 amu
      c. 2(14.01 amu) + 4(1.01 amu) = 32.06 amu
      d. 2(26.98 amu) + 3(16.00 amu) = 101.96 amu

5.3   a. (55.85 amu) + 3(16.00 amu) + 3(1.01 amu) = 106.88 amu
      b. 2(14.01 amu) + 8(1.01 amu) + (32.07 amu) = 68.17 amu
      c. (137.33 amu) + 4(1.01 amu) + 2(30.97 amu) + 8(16.00 amu) = 331.31 amu
      d. (40.08 amu) + 4(12.01 amu) + 6(1.01 amu) + 4(16.00 amu) = 158.18 amu

5.5.   a. (1.00 mole H_2O) x (6.02 x 10^{23} H_2O molecules /1 mole H_2O) = 6.02 x 10^{23} H_2O molecules
      b. (2.00 moles H_2O) x (6.02 x 10^{23} H_2O molecules /1 mole H_2O) = 12.04 x 10^{23} H_2O molecules
      c. (0.500 mole H_2O) x (6.02 x 10^{23} H_2O molecules /1 mole H_2O) = 3.01 x 10^{23} H_2O molecules
      d. (0.621 mole H_2O) x (6.02 x 10^{23} H_2O molecules /1 mole H_2O) = 3.74 x 10^{23} H_2O molecules

5.7   They are the same.

5.9   a. 1.00 mole CO x (28.01 g CO/1 mole CO) = 28.01 g CO
      b. 1.00 mole CO_2 x (44.01 g CO_2/1 mole CO_2) = 44.01 g CO_2
      c. 1.00 mole NaCl x (58.44 g NaCl/1 mole NaCl) = 58.44 g NaCl
      d. 1.00 mole $C_{12}H_{22}O_{11}$ x (342.34 g $C_{12}H_{22}O_{11}$/1 mole $C_{12}H_{22}O_{11}$) = 342.34 g $C_{12}H_{22}O_{11}$

5.11   a. 5.00 g NH_3 x (1 mole NH_3/17.04 g NH_3) = 0.293 mole NH_3
      b. 5.00 g H_2O_2 x (1 mole H_2O_2/34.02 g H_2O_2) = 0.147 mole H_2O_2
      c. 5.00 g SO_2 x (1 mole SO_2/64.07 g SO_2) = 0.0780 mole SO_2
      d. 5.00 g Zn x (1 mole Zn/65.38 g Zn) = 0.0765 mole Zn

5.13 a. 2.00 moles SO_2 molecules x (1 mole S atoms/1 mole SO_2 molecules) = 2.00 moles S atoms
        2.00 moles SO_2 molecules x (2 moles O atoms/1 mole SO_2 molecules) = 4.00 moles O atoms
     b. 2.00 moles SO_3 molecules x (1 mole S atoms/1 mole SO_3 molecules) = 2.00 moles S atoms
        2.00 moles SO_3 molecules x (3 moles O atoms/1 mole SO_3 molecules) = 6.00 moles O atoms
     c. 3.00 moles NH_3 molecules x (1 mole N atoms/1 mole HN_3 molecules) = 3.00 moles N atoms
        3.00 moles NH_3 molecules x (3 moles H atoms/1 mole HN_3 molecules) = 9.00 moles H atoms
     d. 3.00 moles N_2H_4 molecules x (2 moles N atoms/1 mole N_2H_4 molecules)
        = 6.00 moles N atoms
        3.00 moles N_2H_4 molecules x (4 moles H atoms/1 mole N_2H_4 molecules)
        = 12.0 moles H atoms

5.15   a. 0.250 mole S x (32.07 g S/1 mole S) = 8.02 g S
      b. 0.250 mole SO_2 x (64.07 g SO_2/1 mole SO_2) = 16.0 g SO_2
      c. 0.250 mole SO_3 x (80.07 g SO_3/1 mole SO_3) = 20.0 g SO_3
      d. 0.250 mole H_2SO_4 x (98.09 g H_2SO_4/1 mole H_2SO_4) = 24.5 g H_2SO_4

5.17 a. 10.0 g NH_3 x (1 mole NH_3/17.04 g NH_3) x (1 mole N/1 mole NH_3) x (14.01 g N/1 moleN)
= 8.22 g N

b. 10.0 g N_2O_5 x (1 mole N_2O_5/108.02 g N_2O_5) x (2 moles N/1 mole N_2O_5) x
(14.01 g N/1 mole N) = 2.59 g N

c. 10.0 g N_2H_4 x (1 mole N_2H_4/32.06 g N_2H_4) x (2 moles N/1 mole N_2H_4) x
14.01 g N/1 mole N) = 8.74 g N

d. 10.0 g $Al(NO_3)_3$ x [1 mole $Al(NO_3)_3$/213.01 g $Al(NO_3)_3$] x [3 moles N/1 mole $Al(NO_3)_3$] x
(14.01 g N/1 mole N) = 1.97 g N

5.19 a. balanced   b. balanced   c. not balanced   d. balanced

5.21 a. (s) means solid, (g) means gas
b. (g) means gas, (l) means liquid, (aq) means aqueous

5.23 a. $BaCl_2$ + Na_2S $\longrightarrow$ BaS + 2 NaCl

b. Mg + 2 HBr $\longrightarrow$ $MgBr_2$ + H_2

c. 2 Co + 3 $HgCl_2$ $\longrightarrow$ 2 $CoCl_3$ + 3 Hg

d. 2 Na + 2 H_2O $\longrightarrow$ 2 NaOH + H_2

5.25 a. 1 molecule N_2 reacts with 3 molecules H_2 to produce 2 molecules NH_3
1 mole N_2 reacts with 3 moles H_2 to produce 2 moles NH_3

b. 1 molecule CH_4 reacts with 2 molecules O_2 to produce 1 molecule CO_2 and
2 molecules H_2O
1 mole CH_4 reacts with 2 moles O_2 to produce 1 mole CO_2 and 2 moles H_2O

5.27 a. not consistent   b. consistent   c. consistent   d. not consistent

5.29 a. 40.0 g H_2S x (1 mole H_2S/34.09 g H_2S) x (1 mole H_2O_2/1 mole H_2S) x
(34.02 g H_2O_2/1 mole H_2O_2) = 39.9 g H_2O_2

b. 40.0 g H_2S x (1 mole H_2S/34.09 g H_2S) x 1 mole S/1 mole H_2S) x (32.07 g S/1 mole S)
= 37.6 g S

c. 10.0 g H_2O_2 x (1 mole H_2O_2/34.02 g H_2O_2) x (1 mole H_2S/1 mole H_2O_2) = 0.294 mole H_2S

d. 3.40 moles S x (2 moles H_2O/1 mole S) = 6.80 moles H_2O

5.31 3(12.01 amu) + y(1.01 amu) + (32.07 amu) = 76.18 amu; y = 8

5.33 a. 1.00 mole Au; the atomic mass of Au is greater than the atomic mass of Ag
b. 1.00 mole S; the atomic mass of S is greater than the atomic mass of C
c. 1.00 mole Cl_2; Cl_2 has a mass double that of Cl
d. 6.02 x 10^{23} atoms Ne; 1 mole of Ne (6.02 x 10^{23} atoms) has a mass of 20.18 g

5.35 carbon balance: 2(C in butyne) = 8(1); C = 4
hydrogen balance: 2(H in butyne) = 6(2); H = 6
oxygen balance: 2(O in butyne) + 11(2) = 8(2) + 6(1); O = 0
chemical formula of butyne = C_4H_6

5.37 false, you do not divide by the number of atoms present

5.39 true

5.41   true

5.43   true

5.45   false, the total number of atoms on each side of the equation must be equal

5.47   true

5.49   true

5.51   false, 6.02×10^{23} is the number of atoms or molecules in a mole rather than the mass of a mole

5.53   b

5.55   c

5.57   c

5.59   a

5.61   c

Answers to Questions and Problems

6.1   a. kinetic energy   b. potential energy

6.3   a. tend to cause disorder   b. tend to cause order

6.5   through inelastic collisions

6.7   a. A liquid has a definite volume because of significant cohesive forces; a gas has an indefinite volume because of the lack of significant cohesive forces.

      b. The particles in solids and liquids are already very close together; there is very little empty space in a solid or liquid.

6.9   a. 735 mm Hg x (1 atm/760 mm Hg) = 0.967 atm

      b. 0.530 atm x (760 mm Hg/1 atm) = 403 mm Hg

      c. 535 mm Hg x (1 torr/1 mm Hg) = 535 torr

      d. 12.0 psi x (1 atm/14.7 psi) = 0.816 atm

6.11   As one variable increases there is a proportional decrease in the other variable.

6.13   a. $P_2 = P_1V_1/V_2 = (2.0 \text{ atm})(2.0 \text{ L})/(4.0 \text{ L}) = 1.0 \text{ atm}$

      b. $V_2 = P_1V_1/P_2 = (2.0 \text{ atm})(2.0 \text{ L})/(7.0 \text{ atm}) = 0.57 \text{ L}$

      c. $P_1 = P_2V_2/V_1 = (5.0 \text{ atm})(8.0 \text{ L})/(2.0 \text{ L}) = 20 \text{ atm}$

      d. $V_1 = P_2V_2/P_1 = (1.0 \text{ atm})(4.0 \text{ L})/(2.0 \text{ atm}) = 2.0 \text{ L}$

6.15 $V_2 = P_1V_1/P_2 = (655 \text{ mm Hg})(3.0 \text{ L})/(725 \text{ mm Hg}) = 2.71 \text{ L}$

6.17   As one variable increases there is a proportional increase in the other variable.

6.19   a. $T_2 = V_2T_1/V_1 = (4.00 \text{ L})(600 \text{ K})/(2.00 \text{ L}) = 1200 \text{ K} = 927°C$

      b. $V_2 = V_1T_2/T_1 = (2.00 \text{ L})(300 \text{ K})/(600 \text{ K}) = 1.00 \text{ L}$

      c. $T_1 = V_1T_2/V_2 = (2.00 \text{ L})(300 \text{ K})/(8.00 \text{ L}) = 75.0 \text{ K} = -198°C$

      d. $V_1 = V_2T_1/T_2 = (4.00 \text{ L})(400 \text{ K})/(200 \text{ K}) = 8.00 \text{ L}$

6.21 $T_2 = V_2T_1/V_1 = (525 \text{ mL})(298 \text{ K})/(375 \text{ mL}) = 417 \text{ K} = 144°C$

6.23   a. $V_2 = P_1V_1T_2/P_2T_1 = (1.35 \text{ atm})(15.2 \text{ L})(318 \text{ K})/(1.25 \text{ atm})(306 \text{ K}) = 17.1 \text{ L}$

      b. $V_2 = P_1V_1T_2/P_2T_1 = (1.35 \text{ atm})(15.2 \text{ L})(900 \text{ K})/(1.25 \text{ atm})(306 \text{ K}) = 48.3 \text{ L}$

      c. $V_2 = P_1V_1T_2/P_2T_1 = (1.35 \text{ atm})(15.2 \text{ L})(318 \text{ K})/(3.25 \text{ atm})(306 \text{ K}) = 18.6 \text{ L}$

      d. $V_2 = P_1V_1T_2/P_2T_1 = (1.35 \text{ atm})(15.2 \text{ L})(300 \text{ K})/(6.00 \text{ atm})(306 \text{ K}) = 3.35 \text{ L}$

6.25 $V = nRT/P = (0.100 \text{ mole})(0.0821 \text{ atm-L/mole-K})(273 \text{ K})/(2.00 \text{ atm}) = 1.12 \text{ L}$

6.27 $T = PV/nR = (5.23 \text{ atm})(5.23 \text{ L})/(5.23 \text{ moles})(0.0821 \text{ atm-L/mole-K}) = 63.7 \text{ K} = -209°C$

6.29   Each of the individual components of air (O_2, N_2, etc) contributes to the total pressure exerted by the air. The individual contributions to the total pressure are the partial pressures of the components.

6.31 $P_{CO_2} = P_t - P_{O_2} - P_{N_2} - P_{Ar} = (623 - 125 - 175 - 225)$ mm Hg $= 98$ mm Hg

6.33   a. endothermic   b. endothermic   c. exothermic   d. exothermic

6.35   a. evaporation   b. sublimation   c. melting   d. deposition

6.37   amount of liquid decreases, temperature of liquid decreases

6.39   Gaseous molecules of a substance at a temperature and pressure at which we ordinarily would think of the substance as a liquid or solid are referred to as a vapor.

6.41   a. true   b. true   c. true   d. true

6.43   a. At a given temperature, liquids that have strong attractive forces between molecules have lower vapor pressures than liquids that have weak attractive forces between molecules.

        b. At a higher temperature, more molecules have the minimum energy needed to overcome the attractive forces that prevent escape from the liquid.

6.45   a. true   b. false, reduced pressure produces a lower boiling point
        c. true   d. false, pressure is always 1 atmosphere for a normal boiling point

6.47   a. occurs only between polar molecules
        b. occurs between all molecules (polar and nonpolar)
        c. occurs only between molecules that contain hydrogen bonded to a small, very electronegative element

6.49   Hydrogen bonds are much stronger than dipole-dipole interactions.

6.51   Intermolecular forces are only about one-tenth as strong as intramolecular forces.

6.53   a. boils   b. does not boil   c. does not boil   d. does not boil

6.55   a. $P_2 = P_1V_1/V_2 = (1.25 \text{ atm})(575 \text{ mL})/(825 \text{ mL}) = 0.871$ atm
        b. $P_2 = P_1V_1T_2/V_2T_1 = (1.25 \text{ atm})(575 \text{ mL})(448 \text{ K})/(825 \text{ mL})(398 \text{ K}) = 0.981$ atm
        c. $T_2 = V_2T_1/V_1 = (825 \text{ mL})(398 \text{ K})/(575 \text{ mL}) = 571 \text{ K} = 298°C$
        d. $T_2 = P_2V_2T_1/P_1V_1 = (1.75 \text{ atm})(825 \text{ mL})(398 \text{ K})/(1.25 \text{ atm})(575 \text{ mL}) = 799 \text{ K} = 526°C$

6.57   a. PBr_3 b. PI_3 c. PI_3 d. PI_3

6.59   a. not possible   b. possible   c. possible   d. not possible

6.61   true

6.63   false, volume of the gas decreases

6.65   true

6.67   false, the rates are equal at some nonzero value

6.69   false, a normal boiling point involves a pressure of 1 atm rather than an elevation of sea level

6.71   false, intermolecular forces are always weaker than intramolecular forces

6.73   false, strong intermolecular forces result in low vapor pressures

6.75   d

6.77   b

6.79   a

6.81   c

6.83   b

Answers to Questions and Problems

7.1   a. true   b. true   c. true   d. false, solvent and solute interact physically

7.3   Mixtures are of two types: homogeneous and heterogeneous. Homogeneous mixtures are solutions; heterogeneous mixtures are not solutions.

7.5   a. saturated   b. unsaturated
         c. unsaturated (200 g of H_2O)   d. saturated (50 g of H_2O)

7.7   a. aqueous   b. nonaqueous   c. aqueous   d. nonaqueous

7.9   a. decrease   b. increase   c. increase   d. increase

7.11   a. soluble with exceptions   b. soluble
         c. insoluble with exceptions   d. soluble

7.13   a. [6.50 g NaCl/(85.0 + 6.50) g solution] x 100 = 7.10%(m/m)
        b. [2.31 g LiBr/(35.0 + 2.31) g solution] x 100 = 6.19%(m/m)
        c. [12.5 g KNO_3/(125 + 12.5) g solution] x 100 = 9.06%(m/m)
        d. [0.0032 g NaOH/(1.2 + 0.0032) g solution] x 100 = 0.27%(m/m)

7.15   a. (20.0 mL methyl alcohol/475 mL solution) x 100 = 4.21%(v/v)
        b. (4.00 mL bromine/87.0 mL solution) x 100 = 4.60%(v/v)

7.17   a. (5.0 g $MgCl_2$/250 mL solution) x 100 = 2.0%(m/v)
        b. (85 g $MgCl_2$/580 mL solution) x 100 = 15%(m/v)

7.19   a. 3.00 moles KNO_3/0.50 L solution = 6.0 M KNO_3

        b. 12.5 g $C_{12}H_{22}O_{11}$ x (1 mole $C_{12}H_{22}O_{11}$/342.34 g $C_{12}H_{22}O_{11}$) = 0.0365 mole $C_{12}H_{22}O_{11}$
           80.0 mL solution = 0.0800 L solution
           0.0365 mole $C_{12}H_{22}O_{11}$/0.0800 L solution = 0.456 M $C_{12}H_{22}O_{11}$

        c. 25.0 g NaCl x (1 mole NaCl/58.44 g NaCl) = 0.428 mole NaCl
           1250 mL solution = 1.250 L solution
           0.428 mole NaCl/1.250 L solution = 0.342 M NaCl

        d. 0.00125 mole $NaHCO_3$/0.00250 L solution = 0.500 M $NaHCO_3$

7.21   a. 0.220 M NaCl x (25.0 mL/30.0 mL) = 0.183 M NaCl
        b. 0.220 M NaCl x (25.0 mL/75.0 mL) = 0.0733 M NaCl
        c. 0.220 M NaCl x (25.0 mL/457 mL) = 0.0120 M NaCl
        d. 2.00 L = 2000 mL (three significant figures)
           0.220 M NaCl x (25.0 mL/2000 mL) = 0.00275 M NaCl

7.23 a. 5.0 M NaCl x [30.0 mL/(30.0 + 20.0) mL] = 3.0 M NaCl

b. 5.0 M $AgNO_3$ x [30.0 mL/(30.0 + 20.0) mL] = 3.0 M $AgNO_3$

c. 7.5 M NaCl x [30.0 mL/(30.0 + 20.0) mL] = 4.5 M NaCl

d. 2.0 M $AgNO_3$ x [60.0 mL/(60.0 + 20.0) mL] = 1.5 M $AgNO_3$

7.25 The boiling point increases and the freezing point decreases.

7.27 It is lower.

7.29 a. the same as   b. greater than   c. the same as   d. greater than

7.31 a. hemolyze   b. remain unaffected   c. hemolyze   d. crenate

7.33 In osmosis only solvent passes through the membrane. In dialysis both solvent and small
      solute particles pass through the membrane.

7.35 a. like solubility; both soluble   b. unlike solubility
      c. unlike solubility   d. unlike solubility

7.37 3.50 qt solution x (2.00 qt H_2O/100 qt solution) = 0.0700 qt H_2O

7.39 a. 0.400 M K_2SO_4 x (2212 mL/1875 mL) = 0.472 M K_2SO_4

b. 1.25 L = 1250 mL

0.400 M K_2SO_4 x (2212 mL/1250 mL) = 0.708 M K_2SO_4

c. 0.400 M K_2SO_4 x (2212 mL/853 mL) = 1.04 M K_2SO_4

d. 0.400 M K_2SO_4 x (2212 mL/553 mL) = 1.60 M K_2SO_4

7.41 a. 8.00 g NaCl x (1 mole NaCl/58.44 g NaCl) = 0.137 mole NaCl

375 mL solution = 0.375 L solution

0.137 mole NaCl/0.375 L solution = 0.365 M NaCl = 0.730 osmol NaCl ($i = 2$)

4.00 g NaBr x (1 mole NaBr/102.89 g NaBr) = 0.0389 mole NaBr

155 mL solution = 0.155 L solution

0.0389 mole NaBr/0.155 L solution = 0.251 M NaBr = 0.502 osmol NaBr ($i = 2$)

NaCl (0.730 osmol) has a greater osmotic pressure than NaBr (0.502 osmol)

b. 6.00 g NaCl x (1 mole NaCl/58.44 g NaCl) = 0.103 mole NaCl

375 mL solution = 0.375 L solution

0.103 mole NaCl/0.375 L solution = 0.275 M NaCl = 0.550 osmol NaCl ($i = 2$)

6.00 g $MgCl_2$ x (1 mole $MgCl_2$/95.20 g $MgCl_2$) = 0.0630 mole $MgCl_2$

225 mL solution = 0.225 L solution

0.0630 mole $MgCl_2$/0.225 L solution = 0.280 M $MgCl_2$ = 0.840 osmol $MgCl_2$ ($i = 3$)

$MgCl_2$ (0.840 osmol) has a greater osmotic pressure than NaCl (0.550 osmol)

7.43 false, a concentrated solution does not have to be a saturated solution

7.45 true

7.47 false, "like" refers to polarity

7.49   true

7.51   true

7.53   true

7.55   true

7.57   c

7.59   c

7.61   a

7.63   c

7.65   c

Answers to Questions and Problems

8.1   a. $XY \rightarrow X + Y$ b. $X + Y \rightarrow XY$
         c. $AX + BY \rightarrow AY + BX$ d. $X + YZ \rightarrow Y + XZ$

8.3   a. decomposition   b. combination   c. double-replacement   d. combustion

8.5   a. oxidation   b. reduction   c. oxidation   d. oxidation

8.7   a. Li, oxidized; F_2, reduced   b. Fe, oxidized; $CuSO_4$, reduced
         c. C_4H_8, oxidized; O_2, reduced   d. Ca, oxidized; S, reduced

8.9   a. reducing   b. gains

8.11   a. F_2, oxidizing agent; Li, reducing agent   b. Fe_2O_3, oxidizing agent; CO, reducing agent
       c. S, oxidizing agent; Ca, reducing agent   d. Ag_2SO_4, oxidizing agent; Fe, reducing agent

8.13   a. Reactant particles must collide with a certain minimum combined kinetic energy, which is the activation energy.
       b. Orientation relative to one another at the moment of collision is a factor in determining whether a collision produces a reaction.

8.15   a. exothermic   b. energy released

8.17   a. exothermic   b. endothermic   c. endothermic   d. exothermic

8.19   Lowering the temperature decreases the rate of the "spoiling" reaction.

8.21   a. Molecules have more energy and collide more frequently.
       b. A catalyst lowers the activation energy for the reaction.
       c. Physical state affects the freedom of movement of molecules.
       d. At higher concentrations there are more collisions between reactants.

8.23   They are equal.

8.25   a. always applies   b. does not apply   c. always applies   d. always applies

8.27   a. shifts to the left   b. shifts to the left   c. shifts to the left   d. shifts to the left

8.29   a. single-replacement redox   b. combination redox
       c. decomposition redox   d. double-replacement nonredox

8.31   a. H_2O b. Cl_2 c. Cl_2 d. H_2O

8.33   Constant concentrations are not changing, but they do not have to be equal.

8.35   true

8.37   true

8.39 true

8.41 true

8.43 true

8.45 true

8.47 true

8.49 true

8.51 a

8.53 b

8.55 c

8.57 c

8.59 b

14.29 orientation of the –OH group on carbon-4

14.31 a. alpha   b. alpha   c. beta   d. alpha

14.33

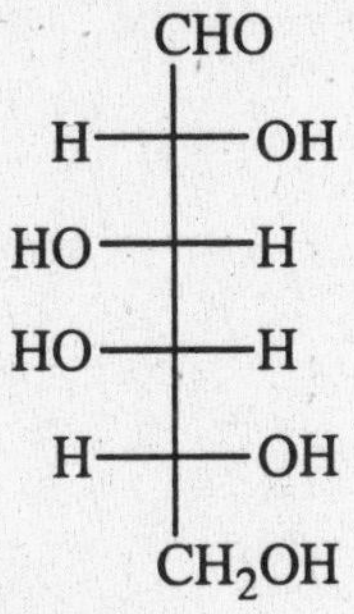

```
        CHO
   H ——————— OH
  HO ——————— H
  HO ——————— H
   H ——————— OH
        CH₂OH
```

14.35 a. yes   b. yes   c. yes   d. yes

14.37 a. yes   b. no   c. yes   d. yes

14.39 a. no   b. no   c. no   d. yes

14.41 a. $\beta(1 \to 4)$ b. $\alpha, \beta(1 \to 2)$ c. $\beta(1 \to 6))$ d. $\alpha(1 \to 4)$

14.43 a. reducing sugar   b. nonreducing sugar   c. reducing sugar   d. reducing sugar

14.45 a. galactose, glucose   b. glucose, fructose   c. glucose, glucose

14.47 a. sucrose   b. lactose   c. maltose

14.49 Animal starch (glycogen) is a more highly branched glucose polymer than plant starch (amylose and amylopectin).

14.51 Bacteria in the intestinal tract produce enzymes that can break the $\beta(1 \to 4)$ glycosidic linkages.

14.53 Cellulose is a polymer of glucose; chitin is a polymer of a glucose derivative.

14.55 a. one   b. two   c. many   d. many

14.57 a. same   b. same   c. same   d. different

14.59 a. yes   b. no   c. no   d. yes

14.61 α-D-galactose is in equilibrium with open-chain-D-galactose which in turn is in equilibrium with the β-D-galactose.

14.63 false, it applies to all monosaccharides and disaccharides

14.65 true

14.67 false, ribose is an aldopentose

14.69 true

14.71 true

14.73 true

14.75 false, glycogen is a highly-branched polymer

14.77 c

14.79 b

14.81 b

14.83 d

14.85 a

Answers to Questions and Problems

15.1 relatively insoluble in water

15.3 nonpolar fatty-acid-containing lipids, polar fatty-acid-containing lipids, and non-fatty-acid-containing lipids

15.5 a. unsaturated   b. polyunsaturated   c. monounsaturated   d. unsaturated

15.7 produces a 30° bend in the carbon chain

15.9 The greater the number of *cis*-double bonds, the lower the melting point.

15.11 three ester functional groups

15.13

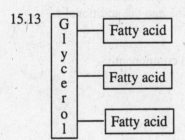

15.15

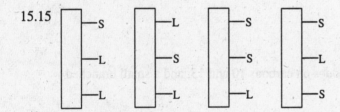

15.17 a. no difference
    b. Fat is a "solid" triacylglycerol.
    c. A mixed triacylglycerol is a triacylglycerol in which at least two different fatty acids are present.
    d. A fat is a "solid" triacylglycerol, and oil is a "liquid" triacylglycerol.

15.19 In general, the fatty acids in oils have a higher degree of unsaturation than do the fatty acids in fats.

15.21 a. glycerol and three fatty acids
    b. a triacylglycerol in which all three fatty acids are saturated

15.23 Antioxidants prevent rancidity; they are more easily oxidized than the fatty acid carbon-carbon double bonds and thus undergo oxidation.

15.25 $CH_2 - CH - CH_2$ $CH_3 - (CH_2)_{14} - COOH$
        | | |
        OH   OH   OH

$CH_3 - (CH_2)_{12} - COOH$ $CH_3 - (CH_2)_7 - CH = CH - (CH_2)_7 - COOH$

15.27 In a triacylglycerol all three of glycerol's –OH groups are esterified with fatty acids; in a phosphoacylglycerol two of glycerol's –OH groups are esterified with fatty acids and the other is esterified with phosphoric acid, which in turn is esterified with an alcohol.

15.29 In a lecithin, choline is esterified to phosphoric acid; in a cephalin an ethanol amine or serine is esterified to phosphoric acid.

15.31 The head is the phosphoric acid-additional component entity and the tails are the two fatty acids.

15.33 The amide linkage involves the sphingosine –NH$_2$ group and the fatty acid; the ester or glycosidic linkage involves the terminal –OH group of sphingosine and the additional component.

15.35 No. Some sphingolipids have a monosaccharide as the additional component.

15.37

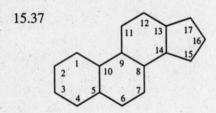

15.39 an –OH group on carbon–3, methyl groups on carbons 10 and 13, and a small branched hydrocarbon chain on carbon–17

15.41 They solubilize lipids.

15.43 sex hormones and adrenocortical hormones

15.45 phosphoacylglycerols and sphingolipids

15.47

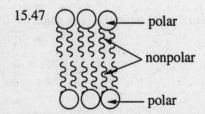

15.49 controls the rigidity of the bilayer

15.51 function as markers important in recognition processes

15.53 a. glycerol-based
       b. glycerol-based
       c. neither glycerol-based nor sphingosine-based
       d. sphingosine-based

15.55 a. triacylglycerol   b. sphingomyelin   c. lecithin   d. cerebroside

15.57 a. 3,0,0   b. 2,1,0   c. 4,0,0   d. 4,0,0

15.59 a. both saturated and unsaturated   b. both saturated and unsaturated
       c. both saturated and unsaturated   d. both saturated and unsaturated

15.61 true

15.63 false, melting point decreases as unsaturation increases

15.65 true

15.67 true

15.69 true

15.71 true

15.73 false, the exterior is polar and the interior is nonpolar

15.75 b

15.77 a

15.79 a

15.81 b

15.83 c

Answers to Questions and Problems

16.1   nitrogen

16.3   amino acids

16.5   nonpolar, polar neutral, polar acidic, and polar basic

16.7   Its side chain is bonded to both the α-carbon atom and the amino-nitrogen atom.

16.9   isoleucine (Ile), asparagine (Asn), glutamine (Gln), and tryptophan (Trp)

16.11 a. no   b. yes   c. yes   d. yes

16.13 It does not possess handedness; no chiral center is present.

16.15 Both an acidic group (–COOH) and a basic group (–NH$_2$) are present.

16.17 a. $H_2N-\overset{\overset{\displaystyle H}{|}}{\underset{\underset{\displaystyle H}{|}}{C}}-COOH$, $H_3\overset{+}{N}-\overset{\overset{\displaystyle H}{|}}{\underset{\underset{\displaystyle H}{|}}{C}}-COO^-$ b. $H_2N-\overset{\overset{\displaystyle H}{|}}{\underset{\underset{\displaystyle CH_3}{|}}{C}}-COOH$, $H_3\overset{+}{N}-\overset{\overset{\displaystyle H}{|}}{\underset{\underset{\displaystyle CH_3}{|}}{C}}-COO^-$

c. $H_2N-\overset{\overset{\displaystyle H}{|}}{\underset{\underset{\underset{\displaystyle CH_3}{|}}{\underset{\displaystyle CH-CH_3}{|}}}{C}}-COOH$, $H_3\overset{+}{N}-\overset{\overset{\displaystyle H}{|}}{\underset{\underset{\underset{\displaystyle CH_3}{|}}{\underset{\displaystyle CH-CH_3}{|}}}{C}}-COO^-$ d. $H_2N-\overset{\overset{\displaystyle H}{|}}{\underset{\underset{\underset{\displaystyle CH_3}{|}}{\underset{\displaystyle CH-OH}{|}}}{C}}-COOH$, $H_3\overset{+}{N}-\overset{\overset{\displaystyle H}{|}}{\underset{\underset{\underset{\displaystyle CH_3}{|}}{\underset{\displaystyle CH-OH}{|}}}{C}}-COO^-$

16.19 a. $H_3\overset{+}{N}-\overset{\overset{\displaystyle H}{|}}{\underset{\underset{\underset{\displaystyle OH}{|}}{\underset{\displaystyle CH_2}{|}}}{C}}-COOH$ b. $H_3\overset{+}{N}-\overset{\overset{\displaystyle H}{|}}{\underset{\underset{\underset{\displaystyle SH}{|}}{\underset{\displaystyle CH_2}{|}}}{C}}-COOH$

c. $H_3\overset{+}{N}-\overset{\overset{\displaystyle H}{|}}{\underset{\underset{\underset{\displaystyle CH_3}{|}}{\underset{\underset{\displaystyle CH-CH_3}{|}}{\underset{\displaystyle CH_2}{|}}}}{C}}-COOH$ d. $H_3\overset{+}{N}-\overset{\overset{\displaystyle H}{|}}{\underset{\underset{\underset{\displaystyle CH_3}{|}}{\underset{\underset{\displaystyle CH_2}{|}}{\underset{\displaystyle CH-CH_3}{|}}}}{C}}-COOH$

16.21 Two –COOH groups are present, and they are deprotonated at different pH values.

16.23 the bond between the carboxyl group of one amino acid and the amino group of another amino acid

16.25 The N-terminal end is the peptide end with a free H_3N^+ group; the C-terminal end is the peptide end with a free COO^- group

16.27 Ser is the N-terminal end in the first dipeptide, and Cys is the N-terminal end in the second dipeptide.

16.29 a. Ser-Val-Cys   b. Asp-Thr-Asn

16.31 a. six   b. six

16.33 primary, secondary, tertiary, quaternary

16.35 They differ in amino acid sequence.

16.37 six (Ala-Ala-Gly-Gly, Gly-Gly-Ala-Ala, Ala-Gly-Ala-Gly, Gly-Ala-Gly-Ala, Ala-Gly-Gly-Ala, Gly-Ala-Ala-Gly)

16.39 carbonyl and amino groups

16.41 Yes, a section of the peptide chain can be α helix and another section β pleated sheet.

16.43 remains the same

16.45 a. noncovalent interaction   b. noncovalent interaction
       c. covalent bond   d. noncovalent interaction

16.47 a. hydrophobic interaction   b. electrostatic interaction
       c. disulfide bond   d. hydrogen bond

16.49 the associations among the separate chains in an oligomeric protein

16.51 a conjugated protein has a non-amino-acid component

16.53 a. lipid   b. carbohydrate   c. phosphate group   d. metal ion

16.55 all levels, (quaternary, tertiary, secondary, and primary)

16.57 yes; both yield Ala and Val

16.59 quaternary, tertiary, and secondary levels

16.61 They have the same primary structure.

16.63 Each biochemical reaction requires a different catalyst.

16.65 Heat can denature the protein present.

16.67 a. oxidation of cytochromes   b. dehydrogenation of alcohols
       c. reduction of L-amino acids   d. hydrolysis of lactose

16.69 a. pyruvate   b. galactose   c. succinate   d. L-amino acids

16.71 A holoenzyme contains an apoenzyme and a cofactor.

16.73 Cofactors provide additional chemically reactive functional groups.

16.75 an intermediate reaction species that is formed when a substrate binds to the active site of the enzyme

16.77 The substrate is the only molecule with a geometry complementary to that of the active site.

16.79 electrostatic interactions, hydrogen bonds, and hydrophobic interactions

16.81 Protein (enzyme) denaturation occurs.

16.83 number of molecules processed per second by one molecule of enzyme functioning under optimum conditions

16.85 a. tertiary   b. tertiary   c. secondary   d. primary

16.87 a. −1   b. −1   c. −4   d. −1

16.89 enzyme + substrate $\rightleftharpoons$ enzyme-substrate complex $\longrightarrow$ enzyme + product

16.91 true

16.93 false, they are left-handed

16.95 true

16.97 false, this is a tertiary structure characteristic

16.99 true

16.101 true

16.103 true

16.105 true

16.107 b

16.109 a

16.111 c

16.113 a

16.115 d

Answers to Questions and Problems

17.1   deoxyribonucleic acid (DNA) and ribonucleic acid (RNA)

17.3   DNA: storage and transfer of genetic information; RNA: synthesis of proteins

17.5   the carbon-2 –OH group in ribose is changed to a –H atom in deoxyribose

17.7   a. pyrimidine   b. pyrimidine   c. purine   d. purine

17.9   a. one   b. four   c. one   d. one

17.11 a. ribose   b. 2′-deoxyribose   c. 2′-deoxyribose   d. ribose

17.13 the alternating chain of sugar and phosphate residues

17.15 the sugar residues

17.17 The 5′-end carries a free phosphate group; the 3′-end carries a free hydroxyl group on carbon-3 of the sugar.

17.19 phosphate and two sugar molecules

17.21 a. Two polynucleotide strands are coiled around each other in a manner somewhat like a spiral staircase.
       b. The outside involves the sugar-phosphate backbones; the inside involves base pairs.

17.23 thymine-thymine-adenine-guanine-cytosine

17.25 T is the 5′-end in T-A; A is the 5′-end in A-T.

17.27 One strand runs in the 5′-to-3′ direction and the other runs in the 3′-to-5′ direction.

17.29 governs the unwinding of the DNA double helix

17.31 Each of the daughter molecules contains one of the parent strands.

17.33 The unwound strands are antiparallel (5′-to-3′ and 3′-to-5′); only the 5′-to-3′ strand can grow continuously.

17.35 transcription

17.37 process by which RNA molecules direct the synthesis of protein molecules

17.39 ribosomal RNA (rRNA), messenger RNA (mRNA), primary transcript RNA (ptRNA), and transfer RNA (tRNA)

17.41 causes the DNA double helix to unwind at a particular location

17.43 ptRNA

17.45   3′ U-A-C-G-A-A-U 5′

17.47   a. a segment of a gene that carries genetic information
        b. a segment of a gene that does not carry genetic information

17.49   a sequence of three nucleotides in an mRNA molecule that codes for a specific amino acid

17.51   There are only 16 two-base code possibilities, and there are 20 standard amino acids.

17.53   a. CUC, CUA, CUG, UUA, UUG
        b. AAC
        c. AGC, UCU, UCC, UCA, UCG
        d. GGU, GGC, GGA

17.55   Met-Lys-Glu-Asp-Leu

17.57   the hairpin loop opposite the open end of the cloverleaf structure

17.59   complementary base pairing

17.61   a. UCU   b. GCA   c. AAA   d. GUU

17.63   a. Ser   b. Ala   c. Lys   d. Val

17.65   a. An amino acid is activated and then forms a complex with a tRNA molecule.
        b. mRNA attaches itself to the P site of a ribosome.
        c. With amino acids in place at the P and A ribosomal sites, the P site amino acid is linked to
            the A site amino acid.
        d. A stop codon terminates the protein synthesis process.

17.67   "A" site

17.69   Gly: GGU, GGC, GGA or GGG
        Ala: GCU, GCC, GCA or GCG
        Cys: UGU or UGC
        Val: GUU, GUC, GUA or GUG
        Tyr: UAU or UAC

17.71   DNA is present as plasmids that replicate independently of the chromosomes; plasmids are
        relatively easily transferred from cell to cell.

17.73   DNA is cleaved in a special manner that produces "sticky ends."

17.75   Recombinant DNA is incorporated in a host cell.

17.77   a. Thymine is a methyl uracil; the methyl group is on carbon-5′.
        b. Adenine is the 6-amino derivative of purine, and guanine is the 2-amino-6-oxo derivative
            of purine.

17.79   a. mRNA   b. tRNA   c. ptRNA   d. tRNA

17.81   A chromosome is a DNA-protein complex; a gene is a segment of a DNA strand.

17.83 a. Single strands of DNA serve as templates; new strands are formed in the 5′-to-3′ direction; base-pairing occurs.

b. replication: entire DNA strand copied; transcription: a segment of a DNA strand copied; repication: new strand remains with the parent strand; transcription: new strand leaves the parent strand; replication: deoxynucleotides are used; transcription: ribonucleotides are used

17.85 false, a nucleotide is a three-subunit molecule

17.87 false, it is a component of deoxyribonucleic acids

17.89 false, the percentages of A and T are equal

17.91 true

17.93 true

17.95 true

17.97 true

17.99 false, they are associated with restriction enzymes

17.101 a

17.103 b

17.105 a

17.107 b

17.109 d

Answers to Questions and Problems

18.1　a. Catabolism: large molecules are broken down into smaller ones; anabolism: small
　　　　 molecules are put together to form larger ones.
　　　 b. Catabolism: energy is usually released; anabolism: energy is usually consumed.

18.3　linear pathway: a series of reactions that generates a final product; cyclic pathway: a series of
　　　 reactions that regenerates the first reactant

18.5　a minute structure within the cell cytoplasm that carries out a specific cellular function

18.7　The outer membrane is permeable to small molecules; the inner membrane is highly
　　　 impermeable to most substances.

18.9　a. adenosine tri phosphate　　　　　　　b. adenosine di phosphate
　　　 c. adenosine mono phosphate　　　　　　d. guanine tri phosphate

18.11　a. $ATP + H_2O \rightarrow ADP + P_i$
　　　　b. $ADP + H_2O \rightarrow AMP + P_i$
　　　　c. $ATP + 2\,H_2O \rightarrow AMP + 2\,P_i$

18.13　FAD (oxidized form) and $FADH_2$ (reduced form)

18.15　flavin subunit

18.17　a.
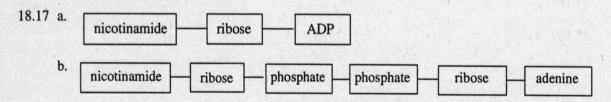
　　　 b.

18.19　a. reduced　　　　　　b. oxidized　　　　　　c. oxidized　　　　　　d. reduced

18.21　sulfhydryl group of 2-aminoethanethiol

18.23　a. stage 1　　　　　　b. stage 3　　　　　　c. stage 4　　　　　　d. stage 2

18.25　a. inside mitochondria　　　　　　　　　b. inside mitochondria
　　　 c. inside mitochondria　　　　　　　　　d. cytoplasm of cell and inside mitochondria

18.27　It is the first product of the cycle.

18.29　exhaled in the process of respiration

18.31　a. two　　　　　　b. one　　　　　　c. three　　　　　　d. one

18.33　a. Steps 3 and 4　　b. Steps 6 and 8　　c. Step 5　　　　d. Step 1

18.35 a. fumarate, malate   b. malate, oxaloacetate
       c. isocitrate, α-ketoglutarate   d. citrate, isocitrate

18.37 NADH and $FADH_2$

18.39 mitochondrion inner membrane

18.41 FeSP, $FADH_2$, cyt c_1, cyt a_3

18.43 a. $FADH_2$, CoQ, $2Fe^{3+}$ b. $FMNH_2$, 2Fe(II)SP, $CoQH_2$

18.45 all three fixed enzyme sites

18.47 10 protons

18.49 ATP synthase

18.51 proton flow through ATP synthase

18.53 2.5 ATP

18.55 They enter the electron transport chain at different stages.

18.57 a. glucose   b. pyruvate

18.59 a. dihydroxyacetone and glyceraldehyde 3-phosphate
       b. It is converted to the one that can be degraded.

18.61 a. 9   b. 2   c. 3   d. 2

18.63 a. glucose 6-phosphate   b. 2-phosphoglycerate
       c. phosphoglyceromutase   d. ADP

18.65 glucose $+ 2\,NAD^+ + 2\,ADP + 2\,P_i \rightarrow 2\,$pyruvate$ + 2\,NADH + 2\,ATP + 2\,H^+ + 2\,H_2O$

18.67 a. pyruvate $+$ CoA $+ NAD^+ \rightarrow$ acetyl CoA $+$ NADH $+ CO_2$
       b. pyruvate $+$ NADH $+ H^+ \rightarrow$ lactate $+ NAD^+$

18.69 NAD^+ is regenerated from NADH.

18.71 Pyruvate is the keto 3-carbon acid, and lactate is the hydroxy 3-carbon acid.

18.73 a. 30 ATP versus 2 ATP   b. 30 ATP versus 2 ATP

18.75 26 ATP

18.77 a. two   b. one

18.79 a. NAD^+ b. CoA, FAD, and NAD^+
       c. ATP, CoA, FAD, NAD^+ d. FAD

18.81 a. a reactant in the citric acid cycle   b. a reactant in glycolysis
       c. a reactant in glycolysis   d. a reactant in the citric acid cycle

18.83 Oxidative phosphorylation produces ATP from ADP and P_i; substrate-level phosphorylation produces ATP from ADP and a phosphate-containing intermediate compound.

18.85 a. ten steps   b. eight steps   c. eight steps   d. four steps

18.87 true

18.89 false, ATP does not contain a B vitamin

18.91 false, most of the intermediates are C_4 molecules

18.93 true

18.95 false, two of the enzymes are mobile-site enzymes

18.97 false, O_2 is a reactant rather than a product

18.99 false, the ETC and oxidative phosphorylation produce significantly more ATP

18.101 c

18.103 a

18.105 d

18.107 c

18.109 c